C J S Petrie

University of Newcastle upon Tyne

Elongational flows
Aspects of the behaviour of model elasticoviscous fluids

π

Pitman

LONDON · SAN FRANCISCO · MELBOURNE

PITMAN PUBLISHING LIMITED
39 Parker Street, London WC2B 5PB

FEARON PITMAN PUBLISHERS INC.
6 Davis Drive, Belmont, California 94002, USA

Associated Companies
Copp Clark Pitman, Toronto
Pitman Publishing New Zealand Ltd, Wellington
Pitman Publishing Pty Ltd, Melbourne

First published 1979

AMS Subject Classifications: (main) 76A10
(subsidiary) 76D99, 34C99, 34B15
34A10, 73B05

Reproduced and printed by photolithography
in Great Britain at Biddles of Guildford.

ISBN 0 273 08406 2

Elongational flows

Aspects of the behaviour of model elasticoviscous fluids

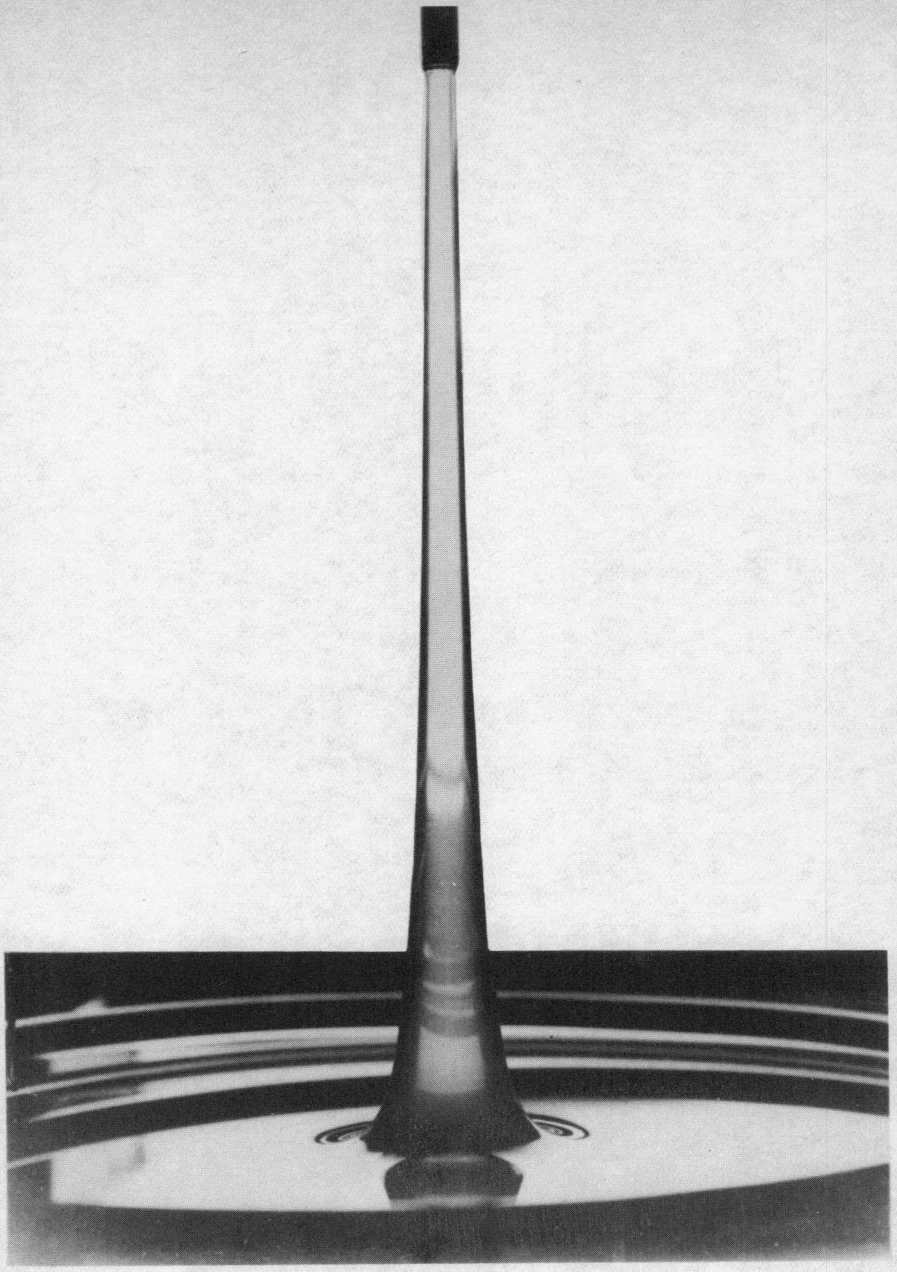

The tubeless siphon (Fano flow)

An ascending column of jet fuel with dissolved polymer. Photograph reproduced by kind permission of S. T. J. Peng: see also reference [P8].

Preface

"My boy, said he, I can work a great deal cheaper than you, because I keep all my goods in the lower story. You have to hoist yours into the upper chambers of the brain, and let them down again to your customers. I take mine in at the level of the ground, and send them off from my doorstep almost without lifting."

Oliver Wendell Holmes.

This book is concerned primarily with its chosen aspect of theoretical rheology and non-Newtonian fluid mechanics. However the subject is not studied for its own sake, and much of the text is devoted to building bridges connecting the main theme with experimental rheology, continuum mechanics, fluid mechanics, physics and engineering. Having referred to these various subjects, I must try to explain what I have in mind by the distinctions implied. Rheology, as the science concerned with the deformation and flow of matter, could be taken to include all the subjects listed except engineering (where the science may be put to use in the design and manufacture of artefacts). The bridge between theory and experiment in rheology concerns the interpretation of observations in terms of a mathematical model of the experiment and in the validation of the model by the testing of its predictions. If the flow is at all complicated, so that we should have a non-trivial problem in the dynamics of a Newtonian liquid, we enter the realm of non-Newtonian fluid mechanics [A18]. Here, in contrast to Newtonian fluid mechanics where we have in essence to solve the Navier-Stokes equations, we do not know what equations we should be solving but are interested in the way the choice of constitutive equation interacts with the dynamics and affects the mathematical problem to be solved. Continuum mechanics (in my usage, which is not uncommon) deals with the general form of the constitutive equation (or rheological equation of state - the relation between stress and deformation for a material) and seeks results which are independent of specific choices, as well as showing rigorously how approximations may lead to specific useable equations. We encounter physics, and need its help, when we begin to try and relate constitutive equations, and the macroscopic behaviour of liquids treated as continua, to the microscopic structure and molecular constitution of liquids.

As we shall see (in chapter 1) there has been an interest in elongational flows (also called extensional flows, or stretching flows) for over seventy years, and this has involved a wide variety of fluids. However the impetus for the recent considerable growth of this topic in connection with rheology comes from work on polymeric liquids (molten polymers and polymer solutions). In view of the economic importance of applications of this work, as well as the physical and mathematical problems of considerable interest and complexity which it offers us, no apology will be made for concentrating largely on the flow of polymers and in problems arising from attempts to model these.

The choice of a physical rather than a mathematical theme for this book means that the mathematical ideas and techniques which we meet vary widely in sophistication. Hence, with a wide variety of readers in mind, I have deliberately given explanations with different amounts of detail throughout the book. The most widespread mathematical topic is nonlinear ordinary differential equations, and the mathematician should find a number of examples inviting, and in some cases deserving, further investigation.

The subject matter falls roughly into three parts. Practical aspects are dealt with in chapter 1, where the history of interest in elongational flow is reviewed and some noteworthy aspects of the behaviour of polymeric liquids are discussed, chapter 3 on experimental techniques for polymers, with an extensive tabulation of published experimental work, and chapter 7 which touches briefly on areas where the theory has been or ought to be used to further our understanding of practical problems. This coverage of the experimental basis of the subject is complemented by the second part which occupies chapters 2 and 4. Here relevant aspects of theoretical rheology and continuum mechanics are presented, and in addition the predictions of material functions such as elongational viscosity are tabulated and discussed for a variety of constitutive equations.

The last section (2.3) of chapter 2 forms a link with the third part in chapters 5 and 6. Here initial-value and boundary-value problems for a number of constitutive equations are formulated and discussed in some detail. This work is in the spirit of non-Newtonian fluid mechanics as outlined above and contains principally material with which I have been closely involved, together with accounts of some work with which I wish I had been associated. Apart from connections made from time to time in the text, the bridge to engineering and applications is probably the most flimsy, and the concluding chapter is no more than an outline of areas of application and of uncertainty. More detailed or specific discussion might have turned this book into a text on polymer processing, and my view is that as well as becoming involved in this, we need to understand the nature of the foundations on which we shall build.

As a research note, rather than a monograph, this book seeks to present up-to-date information in a moderately organised way and to stimulate further research which either builds upon its foundations or investigates them more thoroughly.

Acknowledgements

It is one of the most pleasant priviledges of authorship that one can publicly thank all who have assisted in the task of bringing a book before its readers. I am most grateful to the friends and colleagues who have provided information, comment, encouragement, or various combinations of these, particularly Gianni Astarita, Don Bogue, Bruce Caswell, John Dealy, S. Kase, Guiseppe Marrucci, Kamyar Missaghi, Glen Pearson, S. T. J. Peng, G. V. Vinogradov, Ken Walters, Jim White and Henning Winter. Anthony Pearson and Morton Denn commented extensively on a draft manuscript and I owe them each a particular debt of gratitude. The task of converting manuscript into draft, and draft into final typescript has been cheerfully undertaken by Chris Deighton with assistance from Susan Hurst, and I thank them both. I must also thank Alan Jeffrey for encouraging the idea of writing this book and the publishers, personified by Sue Hemmings, for their kind and patient assistance. The quotations from the writings of Jorge Luis Borges are taken from a collection of stories published in England under the title *Labyrinths* by Penguin Books Ltd., principally from the story *Tlön, Uqbar, Orbis Tertius*. The *Autocrat of the Breakfast-Table* furnished the quotation from Oliver Wendell Holmes and the series *Penguin Modern European Poets* contains the book of poems by Miroslav Holub from which lines from *The Root of the Matter* are taken. Finally I can at last thank my family for suffering an author in the home, and assure them that I look forward as much as they do to the return to a more civilised existence.

Chris Petrie,

November, 1978.

Contents

1 Introduction

"At first it was believed that Tlön was a mere chaos, an irresponsible license of the imagination;"

Jorge Luis Borges.

This book does not advocate a bloody revolution in rheology as a means of toppling shear flows from their pre-eminent position and replacing them by elongational flows. Indeed the number of papers on elongational flows has grown beyond the point where a plaintive note of regret (at the neglect of this worthy field) is appropriate. However a brief discussion of the two classes of flows is necessary, particularly because of the variety of backgrounds from which people come to rheology.

We shall postpone a formal definition of elongational flows, and offer here three examples which will serve as prototypes and will be the focus for discussion in later chapters. These examples are uniaxial stretching, biaxial stretching and spinning. By uniaxial stretching we mean the spatially homogeneous axisymmetric flow of a cylinder of liquid under the action of a tensile force applied at the ends of the cylinder. Biaxial stretching is kinematically the reverse of this, but rather than applying a compressive force to a cylinder we think of applying tensile forces to the edges of a sheet of material. The term spinning is used here in the sense of a spider spinning a web or a silkworm spinning its cocoon, and never in any sense involving rotation or whirling motion. The artificial fibre industry provides many examples from which we can abstract the essentials of a spinning flow, in which liquid flows from an orifice in the form of a narrowing jet or thread which is elongated by a force applied at its further end. This flow, unlike the first two, is not spatially homogeneous, but is steady in the Eulerian sense by which we mean that at any point in space the velocity of the fluid is constant. We note that a particle of the fluid will experience a changing velocity and velocity gradient as it moves along the thread, in other words that the flow is unsteady in the Lagrangean sense.

The reader familiar with the dynamics of viscous fluids will not find the idea of these flows familiar, and he will rightly anticipate considerable experimental difficulty with any but the most viscous liquids. On the other hand to a reader whose background is in solid mechanics the tensile testing of a rod will appear a natural and obvious way of discovering its behaviour under load. Both these readers would, of course, find a simple shearing experiment natural, and so we do have to attempt some justification for an interest in the more difficult experimental arrangement.

There are in fact several reasons for such an interest, which may broadly be attributed to the nonlinear behaviour of polymeric fluids and of the elasticoviscous models we shall be studying here. The important point is that it is not in general possible to predict behaviour in elongation from behaviour in shear, and that most industrially important processes as well as many physically interesting phenomena involve elongation. Furthermore for polymeric liquids there is strong evidence that in flows involving both shear and elongation, the elongation will have a dominant influence. This is most marked for dilute polymer solutions, and has important consequences for rheology viewed as a search for useful and accurate constitutive equations, since it means that elongational flows will provide more sensitive and more relevant tests for discriminating between equations. The same is true for practical tests of processability of polymers, although the dominance of elongational response is perhaps less marked for polymer melts (than for dilute solutions). Finally it is worth remarking on the conceptual simplicity of stretching as opposed to shearing with its associated vorticity or rotation. This simplicity eases the task of calculating kinematic tensors and allows us to exhibit many aspects of elasticoviscous behaviour almost without getting involved in tensor manipulations.

1.1 HISTORICAL NOTE

It is instructive to review the work of Trouton [T24] whose determination of the elongational viscosity of pitch and similar very viscous substances is the earliest such measurement of which this author is aware. The starting-point of Trouton's work is the observation that traditional methods of measuring viscosity "meet with difficulties when it is attempted to apply them to the measurement of the viscosity of bodies such as pitch" [T23]. This earlier paper of Trouton and Andrews has a mention of "methods depending on direct extension and compression" having been used to examine the viscosity of ice, methods which are dismissed since "these apparently did not lead readily to a numerical determination of the coefficient of viscosity". The use of a falling sphere was attempted with shoemaker's wax (being not as viscous as pitch), using X-rays to detect the position of the sphere from time to time. Trouton noted that the descent is irregular, the sphere took a fortnight to travel 1.8 cm and the value obtained for the viscosity varied from 6×10^6 to 23×10^6 poise. The method that Trouton and Andrews found more satisfactory was the torsion of a cylinder of the material under test. They found, unexpectedly, that pitch, glass at elevated temperatures and sodium stearate (soap) all showed viscoelastic behaviour, both in the decay of the rate of deformation from an initially higher value to a steady value under the influence of a constant shear stress, and in the recovery of some deformation when the load was removed.

The classic paper of 1906 [T24] resulted from the further observation [T23] that the rate of strain did not vary linearly with stress in the torsion experiments, and that because the rate of strain varies across the specimen Trouton did not feel that the "exact law connecting the rate of flow with the shearing forces" had been obtained. (The calculation depends on an integration and presupposes the linearity which Trouton did not find.) The results obtained by a variety of elongational flows all showed the same nonlinearity, and the same initial response to loading and recovery on unloading as in the torsional experiments. It is interesting to note that the four methods Trouton used include our three prototypes (if we allow his axial compression to be equivalent to our biaxial stretching). His fourth method was to measure the sagging of a horizontal beam - which requires a calculation like that for the torsion experiment to obtain the viscosity. Trouton's 'spinning' experiment is entitled "Flow of a stream descending under its own weight" and he obtains the viscosity either from measurement of the angle between the tangent to the surface of the stream (or thread) at any point and the vertical, and of the weight of material below that point (found by cutting and weighing), or from a theoretical prediction of the shape of the surface of the thread combined with fitting a curve of the theoretical shape to experimental data to determine a parameter involving the viscosity.

Comparison of the various methods for pitch, pitch and tar mixtures and shoemaker's wax showed good agreement, and comparison with shear viscosity from the torsion experiments confirmed the theoretical prediction that the elongational viscosity (Trouton's "coefficient of viscous traction", λ) is three times the shear viscosity, η. For the mixture with enough tar to allow the descending column method ('spinning') to be used, the comparison was with Poiseuille flow through a tube, and again agreement was adequate. Trouton offers two deductions of the theoretical result ($\lambda = 3\eta$) for which he is most famous; the one "in terms of the more usual analysis of viscous flow", which we shall present below, being relegated to a footnote. His preferred argument, perhaps betraying a background of solid mechanics, is that the tensile stress T may be resolved into three components, two equal shear stresses S and a uniform dilatational stress. Each of these three stresses is equal to T/3, and for an incompressible material the dilatational stress will not cause any flow. Each shear causes a flow $\dot{\phi} = S/\eta$, $\dot{\phi}$ being the rate of rotation of a material line in the plane of the shear and normal to the shear stress. The axial flow is obtained by adding the resolved components of the two shearing flows, and the result $\dot{\gamma} = \dot{\phi}$ is obtained, leading directly to $T/\dot{\gamma} = 3S/\dot{\phi} = 3\eta$ and hence, from the definition $\lambda = T/\dot{\gamma}$, we have $\lambda = 3\eta$ or, in the notation we shall be using,

$$\eta_T = 3\eta \ . \tag{1.1}$$

(We have used $\dot{\gamma}$ here for the axial strain rate, corresponding to a velocity $\dot{\gamma}z$ in the direction of elongation, z.)

1.1.1 Viscosity determination

After Trouton's work there were a number of papers (some apparently unaware of his work), notably on the measurement of the viscosity of glass. English [E 2], Tammann and Jenckel [T 5] and Lillie [L11] all use the rate of elongation of unit length of a glass rod under the action of unit stress, and they call this the mobility (Fließvermogen). This is used in determining the appropriate temperatures for the working and annealing of glass and in this purely technical application no absolute meaning of mobility is required - experience dictates the mobility required for working a glass. In comparing results from this method (at typical temperatures of 500°C to 700°C, with viscosities of 10^7 to 10^{12} Pa.s.) with results for lower viscosities (at higher temperatures) obtained by a rotating rod viscometer, English [E 3] identifies the reciprocal of the mobility (which is the elongational viscosity) with the viscosity, (obtained in simple shear). Stott [S20] points out the error in this and includes the factor of one third but, largely because of the lack of data obtained by both methods at any temperature, both English and Stott are able to draw smooth curves of viscosity against temperature, and the error persists in some later papers [T5]. English [E4] later makes a partial concession, writing that the viscosity is proportional to the reciprocal of the mobility, but nowhere mentions the factor of one third. Later authors such as Lillie [L11], on whose work is based the standard methods of determining the annealing point (ASTM C336-71) and softening point (ASTM C338-57) of glass, and Boow and Turner [B24], use this factor with no comment, nor even a reference to Trouton. Jenckel [J 4] corrects his earlier omission of the factor of one third [J 3], and he in fact redefines the mobility to include a factor of 3 so that it is the reciprocal of the viscosity. The detection and correction of this error are attributed to Philipoff.

After discussing most of the above, and viscosity determination for other "difficult" materials, Van Nieuwenburg [N 6] says that "Reasoning exclusively from a technical point of view, I consider it most urgent that the technique shall be presented with a complete set of different determination methods ... including compression of cylinders, stretching of threads, and the formulae enabling the technical investigator to calculate not only η but also (the rate of strain and the stress) from the determinations". We shall see that in spite of this urgency, this result has not been obtained in the generality desired, largely because of the elasticoviscous behaviour of many materials, and the dependence of many calculations on the choice of constitutive equations used to represent this behaviour.

Scott Blair, a few years later, provides an extensive review of experimental techniques, including bubble inflation in the baking industry, and uniaxial extension or compression of bitumen, butter, resins, rubbers and metals [S 5, pp.94-96]. He also discusses the problem of maintaining constant stress during a large extension, which becomes important in viscoelastic materials (and makes experimental measurement and subsequent calculations easier for purely viscous materials). Andrade [A13] made the first attempt at this, stretching lead wires under the influence of a specially shaped weight which is partially immersed in water. As the weight becomes immersed further the force applied is reduced at the same time as the area of the wire is reduced. We discuss recent techniques in section 3.1.1 below, and note only here that the use of a suitably shaped cam [C24, D 6] appears to have been rediscovered by Leaderman [L6] and by Cogswell, having been used by Andrade and Chalmers [A14] more than 25 years earlier. Andrade's earlier method [A13] was used in some early work on soft (almost solid) polymers [D1] but is not now used.

Of early work on polymers, we shall discuss fibre formation below, and mention here the small number of experiments involving elongational viscosity measurements on polymers prior to 1965. Jenckel and Uberreiter [J5 , J4] investigate the creep and recovery of polystyrene fibres of different molecular masses and obtained mobility (and hence viscosity) and modulus at various temperatures (below 100°C), noting that the former changes much more with temperature. Wiley [W19] carries out tensile tests on specimens of various polymers at temperatures between 130°C and 200°C and measures the rate of elongation of a 6 cm specimen when it reaches a length of 7 cm, which he divides into the stress to get his elongational viscosity. This clearly gives a reproducible number, but the elongation is at constant force so we do not have a property necessarily associated with steady flow in any sense. In fact the graphs of specimen length against time are of different shapes for different polymers, with the rate increasing in a way which corresponds to constant viscosity for polymethyl methacrylate and polystyrene (the curvature of the graph is "almost entirely due to the decrease in cross-sectional area caused by elongation") and being linear for cellulose acetate, implying that there is "an increase in internal friction during flow". Wiley has in mind technical application of his data, (as with Lillie's work on glass), for example by determining empirically the viscosity at which a forming process works well and hence being able to predict the temperature to use for forming with a new polymer (this being the temperature at which it has the desired viscosity). The recognition that characterisation of a material in elongation rather than shear is appropriate when we are interested in processes involving elongation is important, and is reflected in (for example) the modern "melt strength" test [W24] (see also section 3.3.2).

Kargin and Sogolova [K7 ,K8] determine the elongational viscosity of polyisobutylenes of various molecular masses at various temperatures. They use both a constant force and partial hydrostatic compensation (using a cylinder rather than Andrade's specially shaped weight with its hyperbolic profile) leading to a stress which increases in time more slowly than in constant force stretching. The choice of an appropriate measure of strain, and the decomposition of the deformation into reversible (elastic) and irreversible (viscous) parts occupies their attention, and interest in the latter is also seen in modern practical investigations, although theories to deal with this explicitly in nonlinear viscoelasticity have not been developed (except by Leonov [G17,L 9,L10]). Their main finding is that the apparent viscosity (ratio of tensile stress to rate of stretching) is not constant but increases more or less proportionately with the elongation ratio.

Bueche [B28] also measured the recovery of his samples (of polymethyl methacrylate) on removal of the load, and used total non-recoverable (irreversible) deformation to calculate the (average) viscous flow rate, and hence an elongational viscosity. Differentiation of the graph of sample elongation (less recovery) against time provided a more reliable technique; and with elongations of at most 10% the use of constant force was not thought to be significantly different from constant stress. Bueche uses his results as a test of his molecular theory [B27] which offers a prediction (among other things) of the response of a linear polymer in tensile creep and recovery experiments. The agreement at short times (small elongations) is remarkably good. Muller and co-workers [M35,M36] obtain the elongational viscosity of a number of more or less solid polymers (for elongations of 1% or less) and find that this decreases with increased applied stress. They also [B14] consider compression of cylinders, and do not use Trouton's formula (which they do use in extension and which Trouton uses in both cases) but that of Stefan [S17] which is appropriate to the case of a cylinder whose length is much smaller than its diameter. The flow in this case is essentially shearing flow, as fluid is squeezed out from the narrowing gap between two planes compressing the specimen, and the formula is also obtained by Reynolds [R 5] using his lubrication approximation. In this book we shall be considering elongational flows, either extensions of relatively long specimens where end effects due to clamping may be neglected, or compression (in effect biaxial extension) in which the fluid next to the approaching surfaces is free to move and not constrained as in the apparently similar squeezing flow to which the work of Reynolds and Stefan applies. The two formulae are, for a Newtonian liquid of viscosity η in the form of a cylinder of length $h(t)$ and radius $r(t)$,

$$\sigma(t) = 3\eta\dot{h}(t)/h(t) \tag{1.2}$$

due to Trouton and

$$\sigma(t) = 3\eta r(t)^2\dot{h}(t)/\{2h(t)^3\} \tag{1.3}$$

due to Stefan and to Reynolds. Here $\sigma(t)$ is the axial stress, and the dot denotes differentiation with respect to time.

1.1.2 Early theoretical developments

Trouton is frequently given credit just for the empirical verification of his formula (1.1) and not for its theoretical deduction. For example Wiley [W19] states that "Trouton established experimentally that the coefficient of viscosity is equal to one third of the coefficient of viscous traction" and that "Burgers further substantiated this fact theoretically". Now Burgers [B29] made an extremely important contribution to rheology at this time, but as far as the elongational viscosity formula is concerned he has essentially reproduced one of the two derivations offered by Trouton nearly thirty years before [T24]. Burgers does go on to observe that there is no unique way of generalising this result to a non-Newtonian material, where the dependence of viscosity on shear rate may be expressed in a variety of ways in terms of the invariants of the rate of strain tensor. The use of the second invariant alone is offered as a plausible (or practicable) assumption by Burgers, and has been adopted in much subsequent work.

The consequences of allowing a variable viscosity are less interesting and less relevant to the flow of polymeric liquids, than are the consequences of allowing elasticoviscous (or viscoelastic) behaviour, and it is with this latter that we shall mainly be concerned. Burgers is very much aware of this aspect of rheology: "we then must take into consideration the time which is required for the process of change from the original structure into the one appearing under the deformation of the substance, ... When the duration of the experiment is not sufficiently long in order that a definite limiting condition may be reached, the specifications of various data describing the history of the substance will be necessary" [B 29,p.75]. The problems of arranging elongational experiments of sufficiently long duration will become apparent to the reader, and these explain both the interest in elasticoviscous aspects of these flows and the long delay before the fulfilment of Burgers' prophecy: "Hence it appears allowed to say that in general stretching experiments will be equally applicable for the elucidation of the properties of a substance as experiments on the shearing motion."

Another noteworthy contribution in the early days of modern rational continuum mechanics is Reiner's discussion of the coefficient of viscous traction for purely viscous liquids [R3]. He considers the effect of compressibility and volume (or bulk) viscosity on the analysis of stretching flows, something which we shall ignore in assuming that we deal with incompressible materials. In addition he treats the case of a generalised Newtonian liquid and points out how elongation allows us to distinguish liquids where the third invariant of the rate of strain tensor is important. We shall not find this idea, which again deals with a purely viscous fluid, much used in recent years, although a number of authors do consider it to be useful (see, for example, [B1,C22]).

Marshall and Pigford [M15] discuss, as a worked example, the rate of fall of a viscous liquid jet. In this they use (1.1), which they attribute to Burgers, and set up the equations including the effects of gravity and inertia as well as viscosity. Solutions neglecting inertia and neglecting viscosity are seen to be good approximations to the full numerical solution for, respectively, small and large distances from the orifice. An unfortunate error in the equations set up by Marshall and Pigford is pointed out below (section 6.3.1). No further mathematical study of this problem was published for over twenty years.

1.1.3 Fibre formation

Fano [F1] comments, with surprise, on the omission of thread-forming materials from a rheological classification of materials which extends from gases, through viscous and ultraviscous liquids to ductile and finally to brittle materials. The capacity of materials to form threads is widespread among biological fluids as well as polymeric liquids, and attempts to quantify this property of spinnability and to make use of it are of considerable interest. The flows in question have not been fully analysed, or even cast in mathematical terms except in simplified and idealised form and, as we shall see below (section 1.2.3) there are a number of mechanisms associated with the breakage of fibres.

Fano attributes the thread-forming ability of his materials, which included albumen, saliva and an infusion of the leaves of a species of the prickly pear cactus, to the presence of proteins or polysaccharides, and notes that filtration reduces spinnability. His technique [F1] for quantifying this involves what is now known as Fano flow or tubeless siphon flow. A capillary is dipped into the liquid in a beaker, and by connecting a large evacuated vessel to the capillary, liquid is drawn up into it. The level of liquid in the beaker falls, and if a liquid is not spinnable the flow ceases when the level reaches that of the end of the capillary. However a spinnable liquid will continue to flow, and the height of column attainable (between the liquid level in the beaker and the capillary mouth above it) is a measure of spinnability. (See frontispiece.)

Zidan [Z10] showed much more recently how difficult it is to characterise spinnability unequivocally. He repeated Fano's experiment with a polyacrylamide solution and with viscose which was prepared commerically (for artificial fibre manufacture). The polyacrylamide solution is spinnable according to Fano's criterion but the viscose is not, while from the industrial point of view it certainly is spinnable in a commercially important sense. Zidan (whose writing on this point is not altogether clear either to the author or to his German colleague, H. H. Winter) states also that fluids which exhibit normal stress effects (see 1.2.1) may not be spinnable even if they behave as required in Fano's experiment. It is not clear in what sense Zidan is using spinnable in this last comment, nor whether he is implying that in some sense the polyacrylamide solution is not spinnable. Certainly there is room for much discussion on the precise definition of spinnability and other methods have been used to obtain a numerical value to which to give this name.

Tammann and Tampke [T 4] used the technique of withdrawing a glass rod which has been dipped into a liquid (for example molten glucose). The length of thread which is formed is a measure of spinnability, and the square of this length was found to be proportional to the velocity of withdrawal of the rod, and to the viscosity of the material (using the same material at different temperatures). Jochims [J 6] reviewed and developed these techniques and that of Aggazzotti, who measured the length of thread drawn out behind a drop of liquid allowed to fall from a pipette. The main interest is still in physiological fluids, with egg-white (albumen) as a typical example. Work on colloidal systems and polymer solutions followed, most of the detail of which is of interest mainly in the studies of particular systems. There is however further data on the relation between thread length and velocity of pulling, which differs from that of Tammann and Tampke, and shows clearly that different systems behave in different ways [E 5 ,E 6 ,E 7 ,G23,T16,T21,T17]

D'Arcy Thompson [T18,p.12] proposes the term viscidity for what we have here called spinnability and discusses this property and the capillary breakup of threads [T18,pp.61-66]. The viewpoint of a natural historian (protoplasm is the liquid he considers in the main) and the leisurely style make this an account well worth reading, and it is reproduced in full in the abridged edition we have cited. Pearson's criticism [P 5] of the emphasis on surface tension is not wholly justified since for a liquid to be spinnable some mechanism must be delaying the surface tension dominated breakup of very thin threads. The question of the stability of a thread of a spinnable material is not precisely the same question, and we shall have a little more to say on this below (section 1.2.3).

There are practical applications of this scientific study, and as well as the artificial fibre industry, we may instance standards (ASTM-D113-69 and BS 4710) on the ductility of bituminous materials, which date from 1921, and medical and veterinary applications such as the detection of ovulation [S 4 ,C23]. These authors associate the properties of "flow-elasticity" and spinnability, and while it is not established that there is necessarily such an association, it has been remarked on by many of the authors we have cited, and we can do no better than conclude this section by quoting Fano [F 1]. "Spinnable materials are distinguished above all by the fact that, under suitable conditions, they show elasticity in tension to such an extent that one is led to consider them, in respect of some properties at least, to be more like solids than liquids."

1.1.4 Steady elongation

As we have seen, Trouton [T24] used the shape of a jet or stream of liquid falling under its own weight to determine the viscosity of the liquid. Such a flow has obvious application in the formation of artificial fibres and we find Carothers and Hill [C 1] describing both how a filament can be obtained by dipping a glass rod into molten polyester (as in a test for spinnability), and how continuous production of more uniform filaments can be arranged by extruding a viscous solution of the polymer through a spinnerette and allowing the solvent to evaporate from the thread, or by extruding molten polymer and allowing this to solidify. The solidified thread is wound up on a drum, and "it is easy to apply considerable tension to the filaments as they are collected". The stress applied during spinning influences the properties of fibres produced although,more importantly in the fibre industry but of less fluid dynamical interest, subequent drawing of the solid fibre has a much greater effect, involving greater stresses and producing more strongly oriented material.

In a later paper [C 2] Carothers and van Natta describe how molecular mass influences spinnability, the possibility of cold drawing, and the tensile strength of fibres. The latter two properties increase with molecular mass but spinnability increases with molecular mass to a maximum and then at higher molecular masses it decreases again. Erbring [E 6] presents similar data (on spinnability and fibre properties) for solutions of cellulose acetate in a variety of acetone/alcohol mixtures. The necessity of having long chain molecules for fibre spinning is argued by Muller [M34] and Signer [S 8], echoing the remarks of Fano on spinnability (section 1.1.3 above). Further accounts of the historical development of fibre spinning are contained in the review by White [W15] and the monograph by Ziabicki [Z9].

The use of fibre spinning to determine viscosity appears to have been neglected since Trouton until Nitschmann and Schrade [N 7, see also a brief account in N 8]. The attribution of spinnability to the "viscosity anomaly" or "increase in flow resistance with increasing rate of stress" is not in agreement with present ideas (at least if it is interpreted as referring to purely viscous non-Newtonian behaviour, rather than elasticoviscous behaviour). There are, unfortunately, a number of other criticisms which may be levelled against this important pioneering work. The numerical factor of 3 (elongational viscosity divided by shear viscosity) is (again) attributed to Trouton as an empirically determined factor (and indeed Signer in his later reference to this work [S 8] retains an unspecified numerical constant K instead of the factor of 3). Since the analysis of the flow neglects all elastic effects and evaluates the ratio of stress to strain rate at a point on a liquid thread, there is no reason for rejecting the theoretical analysis of Trouton (and of others since then). The expressions for stress and strain rate are, as Vinogradov [V 2] points out, wrong, since Nitschmann and Schrade attempt to obtain the factor of 3 from factors of $\sqrt{3}$ dividing the tensile stress and multiplying the rate of elongational strain.

The comparison in this work of data in shear and elongation for a concentrated polymer solution is important qualitatively, whatever the merits of the method of analysis of the measurements. In addition the "spin balance" is the precursor of a number of more recent experimental arrangements which we mention below (section 3.3). More work in the same laboratory by

Aeschlimann [N 9] is reported by Signer [S 8]. A noteworthy feature of the results is that graphs of the local or instantaneous "viscosity" (tensile stress divided by three times the strain rate at the same point) against stress for various take-up velocities do not superpose, and hence it is deduced that this "viscosity" depends on the previous flow history of the material. We may also give credit to Nitschmann and Schrade [N 7 ,N 8] for remarking on the different effects on molecular orientation of shear and elongation, with a tumbling motion expected in shear, and stable alignment in elongation (see section 4.1.1 below.)

As well as spinning flows, it is interesting to note some early work on jets of elasticoviscous materials, [G 2,G11,G 3] including solutions of some aluminium soaps which are (presumably) similar to liquids whose spinnability was investigated by Erbring [E 5]. Garner, Nissan and Wood [G 2] refer to work done in 1942-46 for the "Petroleum Warfare Department" on sheets of fluids (including Napalm) in the form of conical jets, and Gill and Gavis [G11] refer to a paper by M. Mooney at a 1955 U.S. Army symposium on the "Theory of flame thrower rod stability". This work, which does not contain detailed mechanical information, is of qualitative value, and probably more important in giving the impetus to a good deal of subsequent work on jets and their stability as well as on other aspects of rheology.

1.2 POLYMER MELTS AND SOLUTIONS

1.2.1 Viscous and elastic behaviour

Since interest in elongational flows stems mainly from the properties and uses of polymer melts and solutions, we shall review the rheological behaviour of these polymeric liquids briefly. The first quantitative observation is commonly that polymer melts and concentrated solutions are non-Newtonian in shear; that is to say that the ratio of shear stress to rate of shear strain (or velocity gradient) is not constant. In fact these polymeric liquids are always (as far as this author knows) pseudoplastic or shear-thinning (i.e. the shear stress does not increase as rapidly with increasing shear rate as for a constant viscosity (Newtonian) liquid); we may say that the viscosity in shear is a decreasing function of shear rate. It is also observed that normal stresses are developed in shearing flows, which apparently lead to such phenomena as rod-climbing (the Weissenberg effect - liquid climbs up a rotating rod immersed in it, when the surface of a Newtonian liquid would be depressed). In an incompressible liquid, since an isotropic pressure does not influence deformation or flow of the material, we are concerned only with normal stress differences, and we shall find this both in shearing and elongational flows. There is a considerable literature on the measurement of stresses in shearing flows, to which the book of Walters [W 3] and the review of Pipkin and Tanner [P27] offer an excellent introduction.

Non-zero normal stress differences are associated with elasticity of polymeric liquids, by which we mean the fact that the stress at any instant is determined not only by the rate of strain at that instant but by previous values of the rate of strain for the same bit of material or, equivalently, by the strain relative to previous configurations of the material and by the time which has elapsed during the change of configuration. More direct demonstrations of the elasticity of liquids are available, from simple demonstrations like the bouncing of a ball made of "potty putty" (a silicone polymer available as a toy) which will flow under its own weight if left, to more scientific measurements of recoil or strain recovery when a force causing flow is removed. An allied phenomenon is stress relaxation; when the motion of an incompressible Newtonian liquid ceases the stress becomes isotropic immediately, while with polymeric liquids it takes time for shear stresses and normal stress differences to decay to zero. The other extreme is of course an elastic solid, where the stress will remain constant when motion ceases and will not decay at all.

1.2.2 Time scales and dimensionless numbers

As was suggested earlier the elasticoviscous nature of polymers and the fact that their behaviour is in part liquid-like and in part solid-like is intimately linked with the interesting and important features of behaviour in elongational flows. It becomes apparent that the time-scales of deformations and of flow are of fundamental importance, and this has been recognised from the earliest days of rheology. We must associate a characteristic time with a polymeric liquid (though we shall see later that there is no unique way of doing this - real materials have a spectrum of relaxation times whose relative importance depends on the flow being considered). Then we may define two dimensionless ratios, the Weissenberg number and the Deborah number. Definition of the Weissenberg number is relatively straightforward - it is the product of 'the' characteristic time λ and a characteristic rate of strain $\dot{\gamma}$

$$We = \lambda\dot{\gamma}$$

This can be interpreted as a ratio of elastic or recoverable deformation to viscous (non-recoverable or irreversible) deformation if we think in terms of the most popular simple elasticoviscous model, namely the Maxwell model, discussed below. A small Weissenberg number will correspond to essentially viscous behaviour and a large one to essentially elastic behaviour.

The Deborah number is probably more important, and certainly provokes more discussion. The idea is to define a ratio

$$De = \lambda/\overline{t}$$

where $\overline{t}$ is a characteristic time of the flow, which might be the duration of an experiment, the time for which we watch the flow or the residence time of fluid in a piece of equipment. The question being asked, in looking at the size of the Deborah number is whether memory effects are going to be important, so we are led to the idea of a characteristic time during which the flow changes significantly. This is less easy to deal with simply, but it is an important distinction since in a (materially) steady flow we may have elastic effects and a non-zero Weissenberg number, but wish to distinguish this from effects due to memory in materially unsteady flows which we shall associate with a non-zero Deborah number. The use of a residence time for $\overline{t}$ gives us a minimum value for a Deborah number relevant to the influence of initial conditions or entry effects: $De_{min} = 1$ implies that initial stresses cannot have decayed to zero during flow through a piece of equipment, but $De_{min} \ll 1$ is not sufficient to ensure that initial or upstream flow conditions are irrelevant. This is a topic we shall return to later as it is of great importance in fibre spinning and in many similar flows.

One obvious manifestation of elasticity, and of both Weissenberg and Deborah number effects is die swell (more accurately termed extrudate swell). When a polymeric liquid emerges into the air after flowing through a tube, the emerging jet is observed to have a diameter which may be two or three times as great as that of the tube (or die) from which it emerges. Now this is observed to happen with very long tubes ($De \to 0$) and to depend on shear rate, so that the non-zero Weissenberg number is important. However tube length is also important and with short tubes the swelling ratio can be greatly increased, and can be influenced by such things as the size of reservoir upstream of the tube, clearly showing that there is an important Deborah number effect as well. Extrudate swell is of great practical importance, and must also rank as the major theoretical problem in the mechanics of elasticoviscous liquids awaiting satisfactory resolution. Analysis of this problem is clearly of fundamental importance to any complete analysis of fibre spinning, as we see below (section 6.2.1).

The ideas behind the introduction of these time-scales can be associated with behaviour on a molecular scale, and the kinematical classification of flows as weak or strong becomes closely involved with this (see section 4.1 We remark here only that this brings out a fundamental difference between shearing flows (weak) and elongational flows (strong) since the latter may cause polymer molecules to be extended by a significant amount while the former allow a conformation more like that of a randomly coiled chain. Therefore substantially greater stresses arise in elongational flows than in shearing flows of polymeric liquids and hence elongational behaviour is in many situations the more important.

1.2.3 Instability and fracture

The fact that tensile tests with elongations of several hundred per cent are possible with polymeric liquids is a manifestation of a quite remarkable stability. We have already seen how fibre forming ability (spinnability) has been studied empirically by workers in many fields (section 1.1.3) so this section will be limited to discussing and distinguishing between some basic ideas. It must be said that there is not complete agreement among rheologists, and some subtle points arise, which we shall not explore thoroughly. There are recent discussions of spinnability in review articles by Pearson [P 5], who concentrates on basic theoretical aspects of instability, and by White and Ide [W18, see also I 2, I 3] who, like Petrie and Denn [P14], compare the different mechanisms for thread breakage and review published work. Perhaps the best discussion, by one of the major contributors to our knowledge of this phenomenon, is by Ziabicki [Z 9], who is both careful to keep in view the applications to the manufacture of artificial fibres and thorough in his discussion of the physical processes involved.

We have to consider both the tensile strength of a thread of liquid, which may depend on the precise nature of the flow and its history, and factors which may cause local stress concentrations in the thread and hence breakage before it would be expected on the basis of the overall extension of the thread and the applied force. Theories concerning the cohesive fracture of materials (which is what we term the failure or rupture occurring when the tensile strength of the material is exceeded) were originally statical theories leading to criteria independent of the flow history or, in particular, of the rate of deformation. Such theories are appropriate for elastic materials but not for viscoelastic materials, and Reiner and Freudenthal [R2] developed a 'dynamical theory of strength' based on a maximum stored distortional energy. This depends on the idea that the energy used in deforming an elasticoviscous liquid is in part stored (as elastic or recoverable strain energy) and in part dissipated (in viscous flow), and clearly introduces a rate dependence, since if energy is input so slowly that it may all be dissipated there can be no accumulation of energy to cause rupture. For a simple linear Maxwell model (see below) this leads to failure at a specific stress, and to no limitation on deformation [R4]. Recent experimental evidence from Vinogradov and co-workers [V8,V9] shows that things are not quite so simple, but certainly the Reiner idea offers a valuable first approximation.

It appears that cohesive fracture is important for polymers with higher molecular masses [W18] and at high rates of extension [Z9], corresponding roughly to situations where we should say that the polymers are not spinnable. We should not find this (ultimate) failure surprising, but what is special about the behaviour of polymeric liquids, is that they do not fail at much lower rates of extension. It is therefore to the mechanisms associated with necking (leading to ductile failure) and capillarity (leading to breakup of a thread into droplets) that we should turn, seeking an explanation for the delay in breakup due to the elasticoviscous nature of polymeric liquids. In doing this we must be careful in our choice of words, since we wish to distinguish formally (for theoretical purposes) the marked qualitative difference between elasticoviscous liquids and purely viscous Newtonian liquids, and must remember that it is possible to spin Newtonian liquids, if we are careful. An obvious illustration of this is provided by the drawing of optical fibres which is, according to Geyling [G5], "dominated by viscous stresses, surface tension effects and quenching rates". The latter, that is to say the non-isothermal nature of the process, is certainly highly significant as a stabilizing influence. We must therefore cite laboratory experiments, rather than industrial processes, as evidence that Newtonian liquids can be spun, from Trouton [T24] to more recent work [Z10,W8,M29,B1].

Having said this we must recognise that both capillary breakup and necking are much more important in Newtonian than elasticoviscous liquids, and note evidence [G14,G15] that dissolved polymers can have very large effects on capillary breakup in particular. There is something of a paradox here, superficially at least, since theoretical studies [G14,M30] based on the approach of Rayleigh as modified by Weber to take account of viscosity (see, for example, [W18,G5,C6] for discussion of this early work) predict that viscoelasticity will have a destabilizing effect. Gordon, Yerushalmi and Shinnar [G15] offer a convincing resolution of the paradox, in noting that the initial rate of growth of disturbances for a Separan solution is close to that for a Newtonian liquid, but that later growth in the former case involves the elongation of filaments connecting droplets in the jet, and it is therefore likely that the effect is a manifestation of the very different response to a stretching flow. The paradox is less troublesome when we recall that the classical analysis of capillary instability of a jet does not involve significant elongational flow, and could equally describe an initially stationary column of liquid. The essential mechanism is one of irregularities in a cylindrical surface being magnified, and of a surface tension driven flow opposed by viscosity. Viscoelasticity appears to reduce the viscous opposition to the flow initially, but once a significant amount of stretching has taken place it begins to dominate the whole process.

The idea that the dominant effect is one of much larger resistance to elongational flow than would be found in a Newtonian liquid of the same viscosity in shear can be used also to explain the smaller tendency to neck since necking clearly involves elongation, with an applied force being opposed by the viscosity and elasticity of the material. The analysis of necking for elasticoviscous liquids owes much to the idea of Chang and Lodge [C7] (see also [W18,P14] who seek to define spinnability in terms of the behaviour of the ratio of the areas of two filaments extended by the application of the same force. As we shall see, for a Newtonian liquid the ratio smaller area/larger area tends to zero, while for a rubberlike liquid it tends to a non-zero limiting value. Thus in the former case a neck is drawn down proportionately more than the rest of a filament while in the latter case the increasing resistance to elongation at higher rates of strain prevents this.

In contrast to the desirability of the property of spinnability in the fibre industry, it is of interest to note examples of the reverse. In a sense "sprayability" is the opposite of spinnability, and Elgood and co-workers [E1] record the effect of molecular mass and molecular mass distribution on the sprayability of solutions of polymethyl methacrylate. They note the problems due to "pronounced tendencies to form long filaments" or "cob-webbing". Recent developments in high-speed printing [I1] require reproducible breakup of jets into droplets, and interest in this is reflected in work on the growth of disturbances in jets [R10,B23].

One final important distinction which we must make is between spinnability, or lack of it, and the occurrence of the phenomenon of draw resonance. This is discussed at length by the author [P14], and so here we only note that the sustained oscillation, often of quite large amplitude, known as draw resonance occurs in spinning when the point at which deformation ceases is closely defined, for example by quenching the hot filament suddenly. Also the diameter fluctuations are associated with a fluctuating take-up force (maintaining a constant take-up velocity) and can occur in Newtonian and elasticoviscous fluids. It is this type of instability which a conventional hydrodynamic stability analysis will describe when applied to the two-point boundary-value problem we have in spinning (see section 6.4).

2 Constitutive equations

"The fact that no one believes in the reality of nouns paradoxically causes their number to be unending."

Jorge Luis Borges.

In setting up mathematical models of elongational flows we shall use equations obtained from the physical principles of conservation of mass and of momentum. In addition we need a mathematical description of the response of a material to externally imposed forces or deformations. This will be found in the constitutive equation, or rheological equation of state, of the material which relates the stress at any point to, for example, the strain or rate of strain. We shall describe here some of the simpler equations and present some basic elongational flow results for these and a few other useful or interesting equations. The interested reader is referred to the bibliography below for a discussion of a selection of books on constitutive equations and other topics.

2.1 INCOMPRESSIBLE ELASTICOVISCOUS FLUIDS

We shall confine our attention in this book to incompressible materials. This means that the deformation of the material does not determine the stress tensor uniquely, since any isotropic pressure may be added to it without any effect on the deformation. There is some arbitrariness about the choice of isotropic pressure, but once a constitutive equation is chosen the equations of motion and boundary conditions determine it (uniquely if suitable boundary conditions are prescribed). The basic variable is then the extra-stress tensor, $\underline{p}$, which we define as the tensor given by the constitutive equation (either explicitly or implicitly). Note particularly that $\underline{p}$ need not be traceless (as has been suggested in some early work), nor need it necessarily have a specific interpretation in terms of a molecular model of the material's response, although such an interpretation is often natural and useful.

2.1.1 Linear viscoelasticity

We also restrict discussion to liquids, although recognising that elastico-viscous or viscoelastic behaviour includes many features of solid-like as well as liquid-like behaviour. In Table 2.1 we give six of the simplest models of linear viscoelastic behaviour, using σ and γ for stress and strain without explicitly specifying the flow. In terms of usual notation the equations as written refer either to simple shear or to general flows if σ and γ and their time derivatives (denoted by the dot $\dot{}$) are tensors. In elongation the modulus G would be replaced by Young's modulus, E, and the viscosity η by the elongational viscosity η_T. The initial response is said to be solid-like

	Model	Equation	Initial Response	Ultimate Behaviour
(a)	Hookean	$\sigma = G\gamma$	solid	solid
(b)	Newtonian	$\sigma = \eta\dot{\gamma}$	liquid	liquid
(c)	Kelvin-Voigt	$\sigma = G\gamma+\eta\dot{\gamma}$	liquid	solid
(d)	Maxwell	$\sigma+\lambda\dot{\sigma} = \eta\dot{\gamma}$	solid	liquid
(e)	Jeffreys solid	$\sigma+\lambda\dot{\sigma} = G\gamma+\eta\dot{\gamma}$	solid	solid
(f)	Jeffreys liquid	$\sigma+\lambda\dot{\sigma} = \eta(\dot{\gamma}+\Lambda\ddot{\gamma})$	liquid	liquid

Table 2.1 *Response of simple linear viscoelastic models*

if an instantaneous change in stress corresponds to an instantaneous change in strain. The initial response is said to be liquid-like if an instantaneous change in strain is not possible with a finite stress; then the result of an instantaneous change in stress is an instantaneous change only in the rate of strain, $\dot{\gamma}$, and γ is a continuous function of time. This distinction between solid and liquid, which we shall not use as our main criterion, is due to Bingham [B15] who attributes the alternative distinction, based on the long-term (ultimate) response, to Maxwell. This latter distinction may be put most simply by saying that the material has no "natural" or "preferred" stress-free configuration. The practical consequence of this is that we cannot usefully define an absolute strain and so we exclude the equations in Table 2.1 which use the quantity γ. All the measures of relative strain or relative deformation which we use below are for past configurations, measured relative to the present configuration.

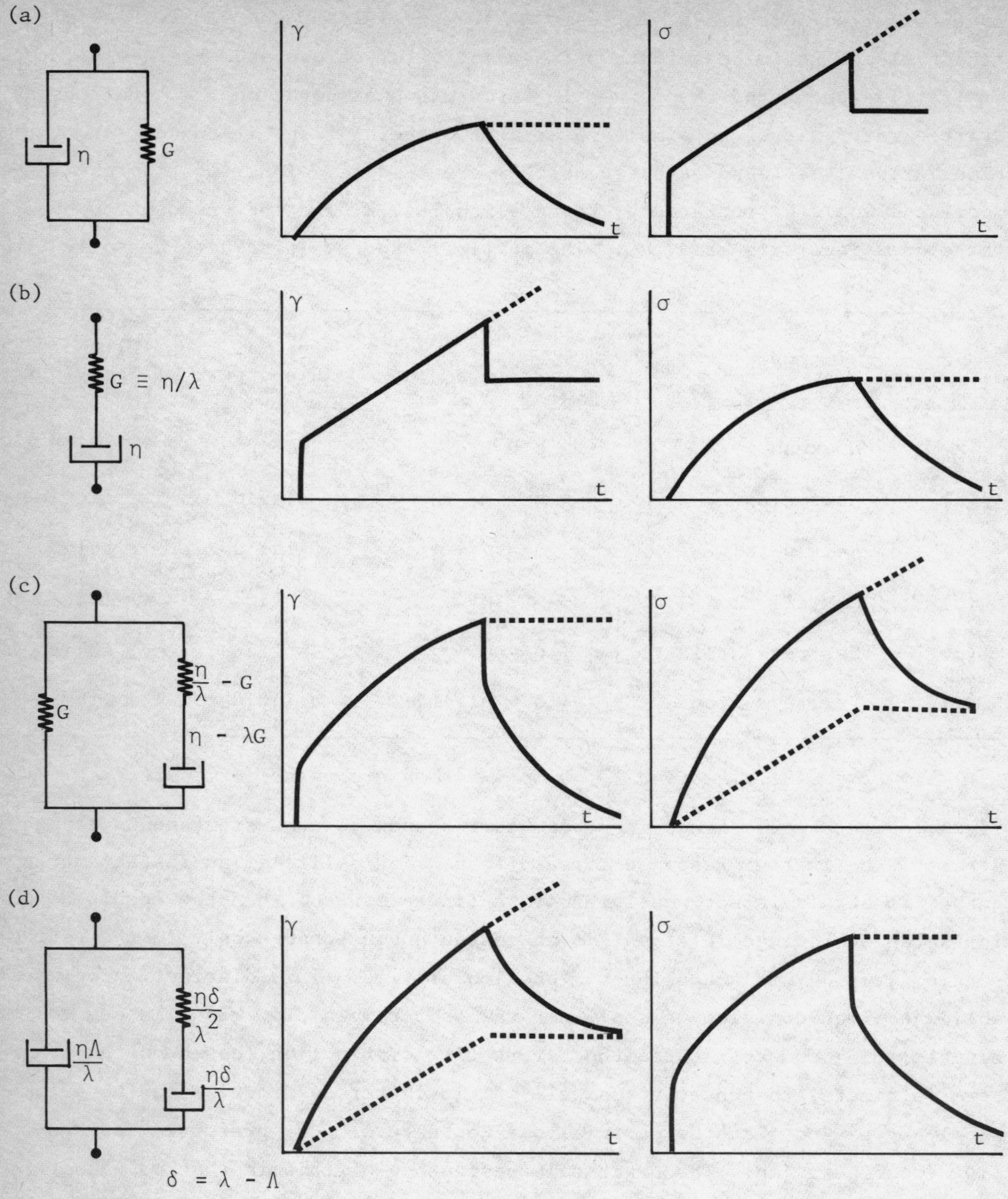

Figure 2.1 Response of simple linear viscoelastic models.

The linear differential equations in Table 2.1 are readily solved for $\sigma(t)$ if $\gamma(t)$ is given, or vice versa, and in Figure 2.1 we show typical graphs (following Giesekus [G10]) for the creep and recovery experiment and for the stressing and relaxation experiment. In the first of these a constant stress is applied for a time (creep) and then removed (recovery) and we record the strain as a function of time. In the second experiment a constant rate of strain is imposed for a time (stressing) and then reduced to zero (relaxation) and the stress is measured. The duality [G10] is apparent from Figure 2.1, in which we also show the mechanical analogues for the four composite models in Table 2.1, in terms of one or two Hookean springs and one or two Newtonian dashpots. Note how the expectation that the moduli of the springs and the viscosities of the dashpots be positive imposes restrictions on the magnitudes of the parameters in the equations (and we also take these to be positive, occasionally with good thermodynamic justification). Such deductions from the mechanical analogues are often helpful, but must not be taken as necessarily true, since the analogies are not physical laws.

The parameter λ, associated with gradual stress relaxation when motion ceases in models (d),(e) and (f), is called a relaxation time, and Λ in (f) or η/G in (c) and (e) is called a retardation time, being associated with delayed or retarded strain recovery when a stress is removed. Noting how the more complex models can be built up by combining Maxwell models in parallel (or Kelvin-Voigt models in series) we may proceed to the general model

$$\sigma = \sum_{k=1}^{M} \sigma_{(k)} \tag{2.1a}$$

$$\sigma_{(k)} + \lambda_k \dot{\sigma}_{(k)} = \lambda_k G_k \dot{\gamma} . \tag{2.1b}$$

If one element is a Hookean spring, $\lambda_k \to \infty$, we have a solid (in long-term response) and similarly if one element is a Newtonian dashpot, for which $\lambda_k \to 0$ with $\lambda_k G_k = \eta_k > 0$, then we have an initially liquidlike response. We shall refer to stresses $\sigma_{(k)}$, or analogous extra-stress tensors $\underline{p}_{(k)}$ as partial stresses corresponding to modes in a relaxation spectrum. Each mode is characterised by a relaxation time λ_k and a modulus G_k, and we shall omit details of continuous spectra which may be thought of as generalisations of the discrete spectra we are considering here. In practice a discrete spectrum is always going to be an adequate approximation.

It is possible to write down integral equations corresponding to these differential equations, for example the Maxwell model

$$\sigma + \lambda\dot{\sigma} = \lambda G\dot{\gamma} \tag{2.2}$$

has the general solution

$$\sigma(t) = \int_{-\infty}^{t} Ge^{-(t-t')/\lambda}\,\dot{\gamma}(t')dt' \tag{2.3}$$

provided that $\sigma(t')\exp(t'/\lambda)\to 0$ as $t'\to -\infty$. Integrating this by parts, and defining the relative strain

$$\gamma_t(t') = \int_{t'}^{t} \dot{\gamma}(t'')dt'' \tag{2.4}$$

gives

$$\sigma(t) = \int_{-\infty}^{t} (G/\lambda)e^{-(t-t')/\lambda}\gamma_t(t')dt' \tag{2.5}$$

For the Jeffreys model,

$$\sigma + \lambda\dot{\sigma} = \eta(\dot{\gamma} + \Lambda\ddot{\gamma}) \tag{2.6}$$

we may use the decomposition of this into a Maxwell and a Newtonian element in parallel to write down immediately

$$\begin{aligned}\sigma(t) &= \sigma_{\text{Maxwell}} + \sigma_{\text{Newtonian}} \\ &= \int_{-\infty}^{t} (\eta/\lambda^2)e^{-(t-t')/\lambda}\gamma_t(t')dt' + \eta(\Lambda/\lambda)\dot{\gamma}(t) .\end{aligned} \tag{2.7}$$

Alternatively use of generalised functions allows us to write an integral like that in (2.5),

$$\sigma(t) = \int_{-\infty}^{t} m(t-t')\,\gamma_t(t')dt' , \tag{2.8a}$$

$$m(t-t') \equiv (\eta/\lambda^2)(1-\Lambda/\lambda)e^{-(t-t')/\lambda} - 2\eta(\Lambda/\lambda)\delta'(t-t') . \tag{2.8b}$$

Here δ' is the derivative of the Dirac delta function with the property

$$\int_{-\infty}^{t} -2\delta'(t-t')f(t')dt' \equiv \dot{f}(t) \tag{2.9}$$

for any integrable function f(t). The function m(t-t') in (2.8a) is referred to as the memory function and is the derivative of the relaxation modulus G(t-t') with respect to t'. For the Jeffreys model the equation corresponding to (2.3) is

$$\sigma(t) = \int_{-\infty}^{t} G(t-t')\dot{\gamma}(t')dt' , \tag{2.10a}$$

$$G(t-t') \equiv (\eta/\lambda)(1-\Lambda/\lambda)e^{-(t-t')/\lambda} + 2\eta(\Lambda/\lambda)\delta(t-t') . \tag{2.10b}$$

(Note that G(t-t') is a function of the indicated argument, while the modulus $G \equiv \eta/\lambda$ and in (2.2),(2.3) and (2.5) is a constant. This is established notation in both cases, and we shall generally avoid confusion by avoiding the use of the relaxation modulus function, using the memory function m(t-t') instead.)

2.1.2 Nonlinear elasticoviscous models

Because of the particular simplicity of elongational flow we could in fact use the above models exactly as given, with σ and $\dot{\gamma}$ standing for any principal stress and the corresponding principal strain rate respectively. In elongational flow the principal axes coincide with each other and with the principal axes of strain, and (2.24) gives us, for finite strains, the logarithmic or Hencky strain (often called the natural measure of strain). For uniform elongation of a specimen with length L(t') at time t' we have

$$\gamma_t(t') = \text{Log}\{L(t)/L(t')\} . \tag{2.11}$$

In more complicated flows, like simple shear, where vorticity or the rotation of principal axes occurs, the equations of linear viscoelasticity no longer apply except as approximations for very small strains. In fact there are many ways of generalising linear viscoelasticity, which correspond both to the use of different nonlinear measures of finite strain, such as (2.11), and to different ways of allowing for rotation. Hence there are at least two Maxwell models corresponding to the strain measure (2.11) [P20], but we shall avoid this problem here.

However in order to use some of the constitutive equations which have been developed, both to keep a link with other work and to allow discussion of shearing and elongation for the same models, we shall summarise the development of a few important equations.

2.1.3 Definitions

A generalisation of the time derivative $\dot{\sigma}$ which corresponds to the natural strain (2.11) in elongation [P15,P17,P18] (but not otherwise [P20]) is the co-rotational or Jaumann derivative. We define this for the extra-stress tensor $\underline{p}$, and for simplicity employ Cartesian components here (and almost everywhere in this book). The definition is

$$\overset{o}{\underline{p}} = \dot{\underline{p}} + \underline{\omega}.\underline{p} - \underline{p}.\underline{\omega} \tag{2.12a}$$

or

$$\overset{o}{p}_{ij} = \dot{p}_{ij} + \omega_{ik}p_{kj} - p_{ik}\omega_{kj} \ . \tag{2.12b}$$

where we adopt the summation convention for repeated suffices, and the definitions

$$\dot{\underline{p}} = \partial\underline{p}/\partial t + (\underline{U}.\nabla)\underline{p} \ , \tag{2.13a}$$

or

$$\dot{p}_{ij} = \partial p_{ij}/\partial t + U_k \partial p_{ij}/\partial X_k \tag{2.13b}$$

and

$$\underline{\omega} = \frac{1}{2}(\underline{L}^T - \underline{L}) \tag{2.14a}$$

or

$$\omega_{ij} = \frac{1}{2}(\partial U_j/\partial X_i - \partial U_i/\partial X_j) \ . \tag{2.14b}$$

The velocity gradient tensor $\underline{L}$ has components $\partial U_i/\partial X_j$, $\underline{L}^T$ is the transpose of $\underline{L}$, U_i are the components of the velocity vector $\underline{U}$ and X_i the coordinates or components of the position vector $\underline{X}$. Here we use the spatial or Eulerian description of motion, treating all the variables as functions of position $\underline{X}$, in some fixed (or laboratory) frame of reference, and time t.

The introduction of $\underline{\omega}$ in (2.12) is sufficient to ensure that there is no unwanted dependence of the material response on motion of the material as a rigid body, or on the observer's frame of reference. Further dependence on this, the anti-symmetric part of $\underline{L}$ is not allowed. The symmetric part of $\underline{L}$ is the rate of strain tensor

$$\underline{e} = \frac{1}{2}(\underline{L} + \underline{L}^T) \tag{2.15a}$$

or

$$e_{ij} = \frac{1}{2}(\partial U_i/\partial X_j + \partial U_j/\partial X_i) \ . \tag{2.15b}$$

This is used to define the upper convected derivative

$$\overset{\nabla}{\underline{p}} = \overset{\circ}{\underline{p}} - (\underline{e}.\underline{p} + \underline{p}.\underline{e}) \tag{2.16a}$$

$$= \dot{\underline{p}} - \underline{L}.\underline{p} - \underline{p}.\underline{L}^T \tag{2.16b}$$

and the lower convected derivative

$$\overset{\Delta}{\underline{p}} = \overset{\circ}{\underline{p}} + (\underline{e}.\underline{p} + \underline{p}.\underline{e}) \tag{2.17a}$$

$$= \dot{\underline{p}} + \underline{L}^T.\underline{p} + \underline{p}.\underline{L} \ . \tag{2.17b}$$

The strain tensors corresponding to these two convected derivatives (as (2.11) corresponds to (2.12) are the strain (due to Signorini)

$$\frac{1}{2}(\underline{C}_t^{-1}(t') - \underline{I}) \tag{2.18a}$$

and the Almansi strain

$$\frac{1}{2}(\underline{I} - \underline{C}_t(t')) \tag{2.19a}$$

where the relative Cauchy deformation tensor $\underline{C}_t$ is defined by

$$\left\{C_t(t')\right\}_{ij} = \frac{\partial X'_k}{\partial X_i}\frac{\partial X'_k}{\partial X_j} \tag{2.19b}$$

and its inverse is the relative Finger deformation tensor

$$\left\{C_t^{-1}(t')\right\}_{ij} = \frac{\partial X_i}{\partial X'_k}\frac{\partial X_j}{\partial X'_k} \tag{2.18b}$$

In these definitions $\underline{X}$ is the position at present time t of a material particle which was at $\underline{X}'$ at the past time t', and $\underline{I}$ is the identity tensor. The sign and the factor of a half in (2.18a) and (2.19a) are chosen to give finite strain tensors which reduce to the convential linear strain for small strains, with a positive strain being associated with stretching of a material line during the motion from $\underline{X}'$ to $\underline{X}$. For the uniform elongation in the X_1 direction of a specimen of length L(t'), the component of $\underline{C}_t$ in the direction of elongation is

$$[C_t(t')]_{11} = L(t')^2/L(t)^2 \ . \tag{2.20}$$

Since we are dealing with incompressible materials the term $\underline{I}$ in the definitions (2.18a) and (2.19a) is not essential; its omission would alter the extra-stress and also (by the same amount) the isotropic pressure.

2.1.4 Rate equations

A convenient generalisation of the Maxwell model is

$$\underline{p} + \lambda\{\overset{o}{\underline{p}} - a(\underline{p}.\underline{e} + \underline{e}.\underline{p})\} = 2\eta\underline{e} \tag{2.21}$$

This includes the upper convected Maxwell model for a = 1, the co-rotational model for a = 0 and the lower convected model for a = -1. Values of a in the range $0 \leq a \leq 1$ are considered worth studying for polymeric liquids. As well as being convenient on purely phenomenological grounds, this equation has some basis in molecular theories. The upper convected model arises both from theories of networks of entangled molecules and from some dilute solution theories, and the parameter a arises in non-affine theories which allow "slip" between molecules and the observable continuum. (See, for example, [B19,G16,J 7,L15,L17,P 20,P 21,P 22].)

This model may be generalised to allow a spectrum of relaxation times,

$$\underline{p} = \sum_{k=1}^{M} \underline{p}_{(k)} \tag{2.22a}$$

$$\underline{p}_{(k)} + \lambda_k \{\overset{o}{\underline{p}}_{(k)} - a(\underline{p}_{(k)}.\underline{e} + \underline{e}.\underline{p}_{(k)})\} = 2\lambda_k G_k \underline{e} \ , \tag{2.22b}$$

and a retardation time, giving a Jeffreys model,

$$\underline{p} + \lambda\{\overset{o}{\underline{p}} - a(\underline{p}.\underline{e} + \underline{e}.\underline{p})\} = 2\eta[\underline{e} + \Lambda\{\overset{o}{\underline{e}} - 2a\underline{e}.\underline{e}\}]. \tag{2.23}$$

The spectrum may be determined empirically (from small-amplitude oscillatory measurements and linear viscoelasticity theory) or postulated on the basis of some theory, as Spriggs [S16] does with M replaced by infinity,

$$\lambda_k = \lambda k^{-\alpha} \tag{2.24a}$$

and

$$G_k = G \Big/ \sum_{j=1}^{\infty} j^{-\alpha} . \tag{2.24b}$$

The molecular theories of Rouse and Zimm suggest this form with $\alpha = 2$, and we thus have an adjustable spectrum for the price of the one extra parameter. In the Jeffreys model (which can also have a spectrum) we are interested in the range $0 \le \Lambda < \lambda$, with $\Lambda = \lambda$ giving an equation which allows any solution the Newtonian equation allows.

Next we have the Oldroyd eight-constant model. Almost all models with a single relaxation time and a single retardation time are special cases of

$$\begin{aligned} &\underline{p} + \lambda_1 \overset{o}{\underline{p}} - \mu_1(\underline{p}.\underline{e} + \underline{e}.\underline{p}) + \nu_1 \mathrm{tr}(\underline{p}.\underline{e})\underline{I} + \mu_o\ \mathrm{tr}(\underline{p})\underline{e} \\ &\quad = 2\eta_o\{\underline{e} + \lambda_2 \overset{o}{\underline{e}} - 2\mu_2 \underline{e}.\underline{e} + \nu_2 \mathrm{tr}(\underline{e}.\underline{e})\underline{I}\} , \end{aligned} \tag{2.25}$$

first proposed by Oldroyd in 1958 [O1,O2]. This was obtained, as a generalisation of the Jeffreys liquid (as presented for a suspension by Frohlich and Sack) by introducing all properly invariant terms linear in p, bilinear in $\underline{p}$ and $\underline{e}$ and quadratic in $\underline{e}$. A model like the Spriggs six-constant model (i.e. with $\mathrm{tr}(\underline{p}) = 0$), but with one relaxation time is obtained by setting $\nu_1 = 2\mu_1/3$, $\nu_2 = 2\mu_2/3$, and then clearly the μ_o term is irrelevant. It does appear, however, that models with $\mu_o \neq 0$ and $\nu_1 = \nu_2 = 0$ have some useful features, although they may be less convenient to use in elongational flows. Their usefulness has been noted generally in computational solution of problems. The idea of making $\underline{p}$ traceless is now generally discarded since in models based on molecular ideas the naturally arising extra-stress has a trace which may be interpreted in terms of stored elastic energy [A4].

All of the above models have constant parameters and when we attempt to approximate real material behaviour over a wide range of conditions this is often inadequate. A number of constitutive equations have been proposed with variable parameters, both with empirically determined functions and with the use of models of molecular behaviour. We shall describe some of these in section 2.2.4 below.

2.1.5 Integral equations

We may write down equations analogous to the simple linear viscoelastic integrals (2.5),(2.8) and (2.9), and, as we have seen, there is a choice of finite strain measures for this generalisation. The best choice of simple model is the rubberlike liquid constitutive equation

$$\underline{p}(t) = \int_{-\infty}^{t} m(t-t')\, \underline{C}_t^{-1}(t')dt' \tag{2.26a}$$

$$m(t-t') = \sum_{k=1}^{M} (G_k/\lambda_k)\exp\{-(t-t')/\lambda_k\} \tag{2.26b}$$

which is a nonlinear integral equation for a set of Maxwell elements in parallel, corresponding to an equation like (2.1) using the upper convected derivative (2.16). With M = 1 this corresponds to the upper convected Maxwell model (2.21) with $a = 1$, $\lambda = \lambda_1$, $\eta = \lambda_1 G_1$. The addition of a Newtonian element as in (2.7) or (2.8) gives a corresponding Jeffreys model.

The integral equation corresponding to (2.21) for any value of the parameter a has been obtained by Johnson and Segalman [J 7],[P20] but we shall not discuss that here since for elongational flows the phenomenological model of Chang, Bloch and Tschoegl [C10] is equivalent, and is more briefly stated as

$$\underline{p}(t) = \int_{-\infty}^{t} 2m(t-t')[\underline{C}_t^{-a}(t') - \underline{I}]/(2a)dt' \tag{2.27a}$$

$$m(t-t') = (G/\lambda)\exp\{-(t-t')/\lambda\} . \tag{2.27b}$$

We shall use the single relaxation time (2.27b) in place of the spectrum (2.26b) frequently. The limiting case $a \to 0$ gives us back the logarithmic strain (2.11).

A development which has been easier in simple shear is the model of Ward and Jenkins

$$\underline{p}(t) = \int_{-\infty}^{t} [m_1(t-t')\underline{C}_t^{-1}(t') + m_2(t-t')\underline{C}_t(t')]dt' \tag{2.28}$$

which is commonly used with m_1 and m_2 proportional to each other, in the form

$$\underline{p}(t) = \int_{-\infty}^{t} m(t-t')\{(1+\tfrac{1}{2}\varepsilon)\underline{C}_t^{-1}(t') + \tfrac{1}{2}\varepsilon\underline{C}_t(t')\}dt' \tag{2.29}$$

with ε negative being expected. The linear combination of Finger and Cauchy deformation tensors can be replaced by the corresponding linear combination of the Almansi and Green strain tensors,

$$2\underline{\overline{S}}_t(t') = \{(2+\varepsilon)\tfrac{1}{2}(\underline{C}_t^{-1}(t') - \underline{I}) - \varepsilon\tfrac{1}{2}(\underline{I} - \underline{C}_t(t'))\} \tag{2.30}$$

with no change in the rheological behaviour for an incompressible liquid. Because this model still, like the rubberlike liquid, predicts a constant viscosity in simple shear there have been many modifications based on allowing the memory function m(t-t') to depend on the motion, through the invariants of $\underline{C}_t^{-1}$, or $\underline{e}$ or $\underline{p}$.

If the upper convected Maxwell model has a relaxation time λ(t) which may vary with time (because it depends on the motion) and if η = G (t) with G constant, the corresponding integral model [A4] is

$$\underline{p}(t) = \int_{-\infty}^{t} \frac{G}{\lambda(t')} \exp\left(-\int_{t'}^{t} \frac{dt''}{\lambda(t'')}\right) \underline{C}_t^{-1}(t')dt' . \tag{2.31}$$

A model very close to this is due to Meister [M25]

$$\underline{p}(t) = \int_{-\infty}^{t} \frac{G}{\theta} \exp\left(-\int_{t'}^{t} \frac{dt''}{g(t'')}\right)\left(\underline{C}_t^{-1}(t') - \underline{I}\right)dt' \tag{2.32}$$

where

$$\lambda(t'') = \theta/[1 + 2b\theta\surd(II_e(t''))] , \tag{2.32b}$$

G, b and θ are constants and II_e is the second invariant of $\underline{e}$, $II_e \equiv \tfrac{1}{2}tr(\underline{e}.\underline{e})$. The effect of allowing the "front factor" G/λ to be constant or to vary is discussed by Chen and Bogue [C13].

We now list four equations with variable spectra (i.e. memory functions affected by the motion) which are all of the form

$$\underline{p}(t) = \int_{-\infty}^{t} 2m(t-t',\ldots)\, \underline{\overline{S}}_t(t')dt' \tag{2.33}$$

with $\underline{\overline{S}}_t(t')$ defined by (2.30), and the motion-dependence of m is to be specified. All the models involve a sum of terms, as in (2.26b), with λ_k, or G_k, or both, allowed to vary.

The Bird-Carreau model [B16] has

$$m(t-t', II_e(t')) = \sum_{k=1}^{\infty}\left(\frac{G_k}{\lambda_k}\frac{\exp\{-(t-t')/\lambda_k\}}{1+\tfrac{1}{2}\tau_k^2 II_e(t')}\right) \tag{2.34a}$$

with

$$\lambda_k = \lambda\left(\frac{2}{1+k}\right)^{\alpha_2} \tag{2.34b}$$

$$\tau_k = \tau\left(\frac{2}{1+k}\right)^{\alpha_1} \tag{2.34c}$$

$$G_k = \eta_o\tau_k\Big/\left(\lambda_k \sum_{j=1}^{\infty} \tau_j\right) \tag{2.34d}$$

and $II_e(t')$ is the second invariant of the rate of strain tensor evaluated at $t = t'$. This is probably the least successful of the five models given here and has in a sense been superseded by the Carreau model B [C 3] which has (in the formulation of Macdonald [M 1])

$$m(t-t';\{II_e(t'')\}) = \sum_{k=1}^{\infty}\left(\frac{G_k}{\lambda_k} f_k \exp\{-\int_{t'}^{t}(\lambda_k g_k)^{-1}dt''\}\right) \tag{2.35a}$$

with

$$\lambda_k = \lambda\left(\frac{2}{1+k}\right)^{\alpha_2} \tag{2.35b}$$

$$\tau_k = \tau\left(\frac{2}{1+k}\right)^{\alpha_1} \tag{2.35c}$$

$$G_k = \eta_o\Big/\sum_{j=1}^{\infty}\lambda_j \tag{2.35d}$$

$$f_k\{II_e(t')\} = \frac{1+\{\frac{1}{2}\tau_k^2 II_e(t')\}^{\alpha_2/\alpha_1}}{\{1+\frac{1}{2}\lambda_k^2 II_e(t')\}^{2R}} \tag{2.35e}$$

$$g_k\{II_e(t'')\} = \frac{\{1+\frac{1}{2}\lambda_k^2 II_e(t'')\}^{R}}{1+\{\frac{1}{2}\tau_k^2 II_e(t'')\}^{\alpha_2/\alpha_1}} \tag{2.35f}$$

with constants α, τ, and R chosen to model the nonlinear behaviour. A model of this sophistication is, not surprisingly, very successful at fitting steady shear flow data and moderately successful with transient data also, but will not be easy to use in complex flows. (Indeed the complexity of equations (2.35a-f) has discouraged any efforts in this direction to date.) We also may note that this model has been still further mofified to the MBC model (Macdonald-Bird-Carreau [M 2] since in spite of its complexity the Carreau model B has deficiencies in predicting some transient stresses in shear, notably the relaxation of normal stresses on cessation of steady shear.

The Bogue-White model has a similar history of development, and in its most popular form [C13] has

$$m\left(t-t'; \overline{II_e^{\frac{1}{2}}(t,t')}\right) = \sum_{k=1}^{n} \frac{G_k}{\lambda_{k(eff)}} \exp\{-(t-t')/\lambda_{k(eff)}\} \tag{2.36a}$$

$$\frac{1}{\lambda_{k(eff)}} = \frac{1}{\lambda_k} + a\overline{II_e^{\frac{1}{2}}(t,t')} \tag{2.36b}$$

$$\overline{II_e^{\frac{1}{2}}(t,t')} = (t-t')^{-1} \int_{t'}^{t} \{II_e(t'')\}^{\frac{1}{2}} dt'' \tag{2.36c}$$

and there is a corresponding model with a continuous spectrum [C13] which has a maximum relaxation time governed by an equation like (2.36b). There is the one nonlinear parameter a, and it is difficult to fit both shear stress and normal stress data (from steady simple shear flow) with a single value of this parameter. The averaging of $II_e^{\frac{1}{2}}$ over the time interval $t' < t'' < t$ may be compared with the similar (more recent) use of the same idea by Agrawal and co-workers [A11].

The network rupture model of Tanner and Simmons [T 7 ,T 8] may be put in the form of (2.33) by defining

$$m(t-t') = \sum_{k=1}^{\infty} \frac{G_k}{\lambda_k} \exp\{-(t-t')/\lambda_k\} \tag{2.37a}$$

for $t-t' < t_R$, and

$$m(t-t') = 0 \tag{2.37b}$$

for $t-t'>t_R$. Alternatively we may use (2.37a),(2.26b) and

$$\underline{p}(t) = \int_{t-t_R}^{t} 2m(t-t')\, \underline{\overline{S}}_t(t')dt' \tag{2.37c}$$

In all of these t_R is the "rupture time" or time at which the "straingth" of the network is exceeded, defined by

$$\mathrm{tr}\{\underline{C}_t^{-1}(t-t_R)\} = B^2 + 3 \qquad (2.37d)$$

(In simple shear, with shear rate κ, this gives us a time $t_R = B/\kappa$.) This equation is thus one in which the memory function depends on the strain or deformation rather than the rate of strain as with the Bird-Carreau model, or a functional of the rate of strain as with the Carreau model B and Bogue-White model.

Our final example is another strain-dependent memory integral equation, the Kaye-BKZ model [K12,B11] which is of the form of (2.28) with

$$m_1(t-t';\, I,II) = \partial U(t-t',\, I,II)/\partial I\ , \qquad (2.38a)$$

$$m_2(t-t';\, I,II) = -\partial U(t-t',\, I,II)/\partial II\ , \qquad (2.38b)$$

and the functional form

$$U(t-t';\, I,II) = \frac{-\alpha'(t-t')}{2}(I-3)^2 - \frac{9\beta'(t-t')}{2}\log\left(\frac{I+II+3}{9}\right)$$

$$- 24\{\beta'(t-t') - c'(t-t')\}\log\left(\frac{I+15}{II+15}\right) - c'(t-t')(I-3) \qquad (2.38c)$$

is suggested by Zapas [Z 4]. Here

$$I = \mathrm{tr}(\underline{C}_t^{-1}(t')),\quad II = \mathrm{tr}(\underline{C}_t(t')) \qquad (2.38d)$$

and the time-dependent terms α',β',c' are derivatives of monotonically decreasing functions of $t-t'$ (and hence are negative, and decreasing in magnitude). Note that Zapas uses $C_{ij}(t,t')$ for the Cauchy deformation tensor of the configuration at t relative to the configuration at t', which is the opposite of our convention (the widely accepted one of taking the present configuration as the reference configuration). Hence his $C_{ij}(t,t')$ is our Finger tensor $\{C_t^{-1}(t')\}_{ij}$ defined by (2.18b). There is some possibility for confusion here (which is evident in some published work).

2.1.6 Other models

There are two models which have not fitted into our scheme of presentation above, but which are probably the two models for which most calculations have been done and boundary-value problems solved. They have been, on the whole, less popular in solving problems for elongational flows than for shear flows. The first of these is the purely viscous power-law liquid

$$\underline{p} = 2C(II_e)^n \underline{e} \tag{2.39}$$

in which $II_e = \frac{1}{2}\text{tr}(\underline{e}.\underline{e})$, and we get a decreasing apparent viscosity in simple shear if $-\frac{1}{2} < n < 0$, and $n = 0$ clearly gives the Newtonian liquid. The model is not good at very low shear rates, when the apparent viscosity tends to infinity rather than to a constant value, but fits other shear viscosity data quite well, sometimes over moderately large ranges of shear rate.

The second omission is the second order fluid

$$\underline{p} = 2\eta\underline{e} - 2\theta\overset{\Delta}{\underline{e}} + 4\chi\underline{e}.\underline{e} \ . \tag{2.40}$$

This is obtained as an approximation to the general simple fluid for slow flows which change smoothly with time. It is not expected to be good for initial value problems which (from the point of view of a material particle) includes all the elongational flows we shall consider. If a slow elongational flow continues for long enough for the influence of the initial conditions to become negligible, we may hope that (2.40) will be a reasonable approximation, but the strain rate must be very small in practice and we cannot expect much more than an indication of how the flow will deviate from that for a Newtonian fluid (which is the ultimate slow flow approximation or "first order fluid"). Note that there is a possibility of confusion with other uses of the term second order fluid since as well as the retarded motion expansion [C30] used to obtain (2.40), it is possible to expand the constitutive functional for a simple fluid in terms of a series of repeated integrals [G20,P25,B18,G12]. To avoid ambiguity (2.40) may be called a second order fluid of differential type (the term fluid of second grade is not preferred as it is also used for non-simple fluids).

2.2 MATERIAL FUNCTIONS

2.2.1 Basic steady flows

In order to calculate a material function, we postulate a uniform steady flow, that is to say a flow where the stress and rate of strain tensors are constants. We discuss later whether such flows may in fact be realised (section 2.2.3) and give a fuller discussion of the basic elongational flows in section 2.3 and chapter 5.

We consider here four simple flows, namely steady simple shear

$$\underline{e} = \tfrac{1}{2}\begin{pmatrix} 0 & \kappa & 0 \\ \kappa & 0 & 0 \\ 0 & 0 & 0 \end{pmatrix} , \quad \underline{C}_t^{-1}(t') = \begin{pmatrix} 1+k^2 & k & 0 \\ k & 1 & 0 \\ 0 & 0 & 1 \end{pmatrix} \qquad (2.41a,b)$$

with constant
with constant shear rate κ and shear $k = \kappa(t-t')$; steady uniaxial elongation

$$\underline{e} = \begin{pmatrix} \dot{\gamma} & 0 & 0 \\ 0 & -\tfrac{1}{2}\dot{\gamma} & 0 \\ 0 & 0 & -\tfrac{1}{2}\dot{\gamma} \end{pmatrix} , \quad \underline{C}_t^{-1}(t') = \begin{pmatrix} \exp 2\gamma & 0 & 0 \\ 0 & \exp(-\gamma) & 0 \\ 0 & 0 & \exp(-\gamma) \end{pmatrix} \qquad (2.42a,b)$$

with constant strain rate $\dot{\gamma}$ and strain $\gamma = \dot{\gamma}(t-t')$; steady equal biaxial elongation

$$\underline{e} = \begin{pmatrix} -2\dot{\Gamma} & 0 & 0 \\ 0 & \dot{\Gamma} & 0 \\ 0 & 0 & \dot{\Gamma} \end{pmatrix} , \quad \underline{C}_t^{-1}(t') = \begin{pmatrix} \exp(-4\Gamma) & 0 & 0 \\ 0 & \exp 2\Gamma & 0 \\ 0 & 0 & \exp 2\Gamma \end{pmatrix} \qquad (2.43a,b)$$

with constant strain rate $\dot{\Gamma}$ and strain $\Gamma = \dot{\Gamma}(t-t')$; and steady pure shear, or strip biaxial extension,

$$\underline{e} = \begin{pmatrix} \dot{\varepsilon} & 0 & 0 \\ 0 & -\dot{\varepsilon} & 0 \\ 0 & 0 & 0 \end{pmatrix} , \quad \underline{C}_t^{-1}(t') = \begin{pmatrix} \exp 2\varepsilon & 0 & 0 \\ 0 & \exp(-2\varepsilon) & 0 \\ 0 & 0 & 1 \end{pmatrix} \qquad (2.44a,b)$$

with constant strain rate $\dot{\varepsilon}$ and strain $\varepsilon = \dot{\varepsilon}(t-t')$. Apart from the first these are irrotational flows, while for steady simple shear (where U $U_1 = \kappa X_2$, $U_2 = U_3 = 0$) we have the vorticity tensor (2.14)

$$\underline{\omega} = \tfrac{1}{2}\begin{pmatrix} 0 & -\kappa & 0 \\ \kappa & 0 & 0 \\ 0 & 0 & 0 \end{pmatrix} . \qquad (2.41c)$$

We define the viscosity

$$\eta(\kappa) = P_{12}/\kappa \qquad (2.45a)$$

and the normal stress functions

$$\Psi_1(\kappa) = N_1/\kappa^2 = (p_{11} - p_{22})/\kappa^2 \tag{2.45b}$$

and

$$\Psi_2(\kappa) = N_2/\kappa^2 = (p_{22} - p_{33})/\kappa^2 \tag{2.45c}$$

in steady simple shear. In steady uniaxial elongation we have the elongational viscosity (or Trouton viscosity)

$$\eta_T(\dot{\gamma}) = (p_{11} - p_{22})/\dot{\gamma} \tag{2.46}$$

and in biaxial elongation we have, respectively

$$\eta_{eb}(\dot{\Gamma}) = (p_{33} - p_{11})/\dot{\Gamma} \tag{2.47}$$

and

$$\eta_{sb}(\dot{\varepsilon}) = (p_{11} - p_{22})/\dot{\varepsilon} \tag{2.48a}$$

In strip biaxial extension we also have the function

$$\eta^x_{sb}(\dot{\varepsilon}) = (p_{33} - p_{22})/\dot{\varepsilon} \tag{2.48b}$$

relating the stress applied to the side of the strip to prevent motion in the X_3-direction to the strain rate. Since it has been shown [D4] that

$$\eta_{eb}(\dot{\Gamma}) = 2\eta_T(-2\dot{\Gamma}) \tag{2.39}$$

we shall not tabulate η_{eb} separately. The similar

$$\eta_{sb}(\dot{\varepsilon}) = \frac{4}{3}\,\eta_T\left(\frac{2\dot{\varepsilon}}{\sqrt{3}}\right) \tag{2.40}$$

is not generally true, but holds only for a Newtonian liquid and a "Generalised Newtonian Fluid" whose viscosity is a function of the second invariant of the rate of strain tensor only. Thus for a viscoelastic fluid we get essentially different information from strip biaxial experiments. Of course we cannot argue that uniaxial and equal biaxial experiments are equivalent since, as Dealy [D 4] points out, one gives us the function $\eta_T(\dot{\gamma})$ for positive arguments only and the other gives the same function for negative arguments only, through (2.39), and there is no a priori reasoning by which we may deduce the one from the other. The elongational viscosity will not generally be either an odd or an even function of $\dot{\gamma}$.

We note that, if we wish to attempt a comparison of these functions, the invariants of the tensors $\underline{e}$ and $\underline{C}_t^{-1}(t')$ are, for the four cases in order, and with $I_e = tr(\underline{e}) \equiv 0$, $II_e = \frac{1}{2}tr(\underline{e}.\underline{e})$, $III_e = det(\underline{e})$, $I = tr(\underline{C}_t^{-1}(t'))$, $II = tr(\underline{C}_t(t'))$, $III = det(\underline{C}_t^{-1}(t')) \equiv 1$:

$$II_e = \kappa^2/4,\ III_e = 0,\ I = II = 3+k^2\ ; \tag{2.49a}$$

$$II_e = 3\dot{\gamma}^2/4,\ III_e = \dot{\gamma}^3/4,\ I = \exp 2\gamma + 2\exp(-\gamma),\ II = \exp(-2\gamma)+2\exp\gamma\ ; \tag{2.49b}$$

$$II_e = 3\dot{\Gamma}^2,\ III_e = -2\dot{\Gamma}^3,\ I = \exp(-4\Gamma)+2\exp 2\Gamma,\ II = \exp 4\Gamma + 2\exp(-2\Gamma)\ ; \tag{2.49c}$$

$$II_e = \dot{\varepsilon}^2,\ III_e = 0,\ I = II = 1 + \exp 2\varepsilon + \exp(-2\varepsilon)\ . \tag{2.49d}$$

In the cases of simple shear and pure shear (strip biaxial extension) we have symmetry arguments from which we deduce that $\eta(\kappa)$, $\Psi_1(\kappa)$, $\Psi_2(\kappa)$ and $\eta_{sb}(\dot{\varepsilon})$ are even functions of their arguments while η_{sb}^x, as well as η_T and η_{eb} are not. The argument for η_{sb} is that by changing the sign of $\dot{\varepsilon}$ we in effect change the roles of the X_1- and X_2-axes, so that the flow is equivalent to strip biaxial extension with the applied force $p_{22}-p_{11}$ in the X_2-direction, and $\{p_{22}(-\dot{\varepsilon})-p_{11}(-\dot{\varepsilon})\}/\{-\dot{\varepsilon}\}$ must be the same as $\{p_{11}(\dot{\varepsilon})-p_{22}(\dot{\varepsilon})\}/\{\dot{\varepsilon}\}$. The same is not necessarily true of the individual components of the extra-stress tensor, p_{ii}, and hence is not necessarily true of $\{p_{33}(\dot{\varepsilon})-p_{22}(\dot{\varepsilon})\}/\{\dot{\varepsilon}\}$. All we can say is that the sideways force should be unchanged, irrespective of whether we have $\dot{\varepsilon}$ or $-\dot{\varepsilon}$, i.e. that

$$p_{33}(\dot{\varepsilon})-p_{22}(\dot{\varepsilon}) = p_{33}(-\dot{\varepsilon})-p_{11}(-\dot{\varepsilon})\ . \tag{2.50a}$$

From this we may deduce that

$$\begin{aligned}\eta_{sb}^x(\dot{\varepsilon}) &= [p_{33}(\dot{\varepsilon})-p_{22}(\dot{\varepsilon})]/\dot{\varepsilon} = -[p_{33}(-\dot{\varepsilon})-p_{11}(-\dot{\varepsilon})]/(-\dot{\varepsilon}) \\ &= \eta_{sb}(-\dot{\varepsilon})-\eta_{sb}^x(-\dot{\varepsilon})\ .\end{aligned} \tag{2.50b}$$

Note, finally, that in calculating steady flow properties we shall obtain the same results with several equations, or at least shall obtain viscosity functions which may be very similar, for models which have different transient responses (and which have in some cases been constructed explicitly to model transient behaviour).

2.2.2 Explicit material functions

In Table 2.2 we show the shearing and elongational viscosity functions for a series of rate equations, all of which are special cases of (2.25) - the Olroyd eight-constant model. The general formulae for (2.25) are, in steady simple shear, [O1]

$$\eta(\kappa) = \eta_o(1 + \sigma_2\kappa^2)/(1 + \sigma_1\kappa^2) \ , \tag{2.51a}$$

$$\Psi_1(\kappa) = 2\lambda_1\eta(\kappa) - 2\lambda_2\eta_o \ . \tag{2.51b}$$

$$\Psi_2(\kappa) = -(\lambda_1 - \mu_1)\eta(\kappa) + (\lambda_2 - \mu_2)\eta_o \tag{2.51c}$$

where

$$\sigma_1 = \lambda_1^2 + \mu_o(\mu_1 - 3\nu_1/2) - \mu_1(\mu_1 - \nu_1) \ , \tag{2.51d}$$

$$\sigma_2 = \lambda_1\lambda_2 + \mu_o(\mu_2 - 3\nu_2/2) - \mu_1(\mu_2 - \nu_2) \tag{2.51e}$$

and in elongation

$$\eta_T(\dot{\gamma}) = 3\eta_o \frac{\{1-\mu_2\dot{\gamma}+(3\nu_2-2\mu_2)(\mu_1-3\mu_o/2)\dot{\gamma}^2}{\{1-\mu_1\dot{\gamma}+(3\nu_1-2\mu_1)(\mu_1-3\mu_o/2)\dot{\gamma}^2} \ , \tag{2.52}$$

$$\eta_{sb}(\dot{\varepsilon}) = 4\eta_o \frac{\{1+4(\sigma_2-\lambda_1\lambda_2)\dot{\varepsilon}^2}{\{1+4(\sigma_1-\lambda_1{}^2)\dot{\varepsilon}^2} \ , \tag{2.53a}$$

$$\eta^x_{sb}(\dot{\varepsilon}) = 2\eta_o \frac{\{1-2(\mu_1-\mu_2)\dot{\varepsilon}+4(\sigma_2-\lambda_1\lambda_2)\dot{\varepsilon}^2+4(\mu_1\nu_2-\mu_2\nu_1)(3\mu_o-2\mu_1)\dot{\varepsilon}^3\}}{\{1+4(\sigma_1-\lambda_1{}^2)\dot{\varepsilon}^2\}} \tag{2.53b}$$

For the integral models in Table 2.3 we show the effect of the parameter ε of (2.30) with the Ward-Jenkins constant spectrum model, and set $\varepsilon = 0$ for all the other models for simplicity (which of course makes Ψ_2 zero). We take the memory function of the form $(G/\lambda)\exp\{-(t-t')/\lambda\}$ with one relaxation time, for simplicity and for direct comparison with rate equations. (This obviously affects some equations more than others and the full capability of the more sophisticated equations is most easily seen graphically, and requires numerical computation. The references cited above are the best

Model	*Number of constants*	*Defining equation*	$\psi_1(\kappa)$	$\psi_2(\kappa)$
Newtonian	1	-	0	0
Second order fluid	3	(2.40)	2θ	$-(2\theta-\chi)$
Upper convected Maxwell	2	a=1 in (2.21)	$2\lambda\eta$	0
Co-rotational Maxwell	2	a=0 in (2.21)	$\frac{2\lambda\eta}{1+\lambda^2\kappa^2}$	$\frac{-\lambda\eta}{1+\lambda^2\kappa^2}$
Lower convected Maxwell	2	a=-1 in (2.21)	$2\lambda\eta$	$-2\lambda\eta$
Generalized convected Maxwell	3	(2.21)	$\frac{2\lambda\eta}{1+(1-a^2)\lambda^2\kappa^2}$	$\frac{-(1-a)\lambda\eta}{1+(1-a^2)\lambda^2\kappa^2}$
Upper convected Jeffreys	3	a=1 in (2.23)	$2\eta(\lambda-\Lambda)$	0
Co-rotational Jeffreys	3	a=0 in (2.23)	$\frac{2\eta(\lambda-\Lambda)}{1+\lambda^2\kappa^2}$	$\frac{-\eta(\lambda-\Lambda)}{1+\lambda^2\kappa^2}$
Generalized convected Jeffreys	4	(2.23)	$\frac{2\eta(\lambda-\Lambda)}{1+(1-a^2)\lambda^2\kappa^2}$	$\frac{-(1-a)\eta(\lambda-\Lambda)}{1+(1-a^2)\lambda^2\kappa^2}$

Table 2.2 Material functions in steady shearing

source of this information.)

We may obtain approximate values of the material functions for small and large strain rates, and in some cases for small or large values of some of the parameters. For example at large elongation rate the elongational viscosity of the Tanner-Simmons model is given by

$$\eta_T = \frac{G\lambda B}{2\lambda^2\dot{\gamma}^2} + O(\dot{\gamma}^{-3})$$

$\eta(\kappa)$	$\eta_T(\dot\gamma)$	$\eta_{sb}(\dot\varepsilon)$	$\eta^x_{sb}(\dot\varepsilon)$
η	3η	4η	2η
η	$3\eta+3(\chi-\theta)\dot\gamma$	4η	$2\eta-4(\chi-\theta)\dot\varepsilon$
η	$\frac{3\eta}{(1+\lambda\dot\gamma)(1-2\lambda\dot\gamma)}$	$\frac{4\eta}{1-4\lambda^2\dot\varepsilon^2}$	$\frac{2\eta}{1+2\lambda\dot\varepsilon}$
$\frac{\eta}{1+\lambda^2\kappa^2}$	3η	4η	2η
η	$\frac{3\eta}{(1-\lambda\dot\gamma)(1+2\lambda\dot\gamma)}$	$\frac{4\eta}{1-4\lambda^2\dot\varepsilon^2}$	$\frac{2\eta}{1-2\lambda\dot\varepsilon}$
$\frac{\eta}{1+(1-a^2)\lambda^2\kappa^2}$	$\frac{3\eta}{(1+a\lambda\dot\gamma)(1-2a\lambda\dot\gamma)}$	$\frac{4\eta}{1-4a^2\lambda^2\dot\varepsilon^2}$	$\frac{2\eta}{1+2a\lambda\dot\varepsilon}$
η	$3\eta\left(\frac{1-\Lambda\dot\gamma(1+2\lambda\dot\gamma)}{(1+\lambda\dot\gamma)(1-2\lambda\dot\gamma)}\right)$	$4\eta\left(1+\frac{4(\lambda-\Lambda)\lambda\dot\varepsilon^2}{1-4\lambda^2\dot\varepsilon^2}\right)$	$2\eta\left(1-\frac{2(\lambda-\Lambda)\dot\varepsilon}{1+2\lambda\dot\varepsilon}\right)$
$\eta-\frac{(\lambda-\Lambda)\lambda\eta\kappa^2}{1+\lambda^2\kappa^2}$	3η	4η	2η
$\frac{\eta+\eta\lambda\Lambda(1-a^2)\kappa^2}{1+(1-a^2)\lambda^2\kappa^2}$	$3\eta\left(\frac{1-a\Lambda\dot\gamma(1+2a\lambda\dot\gamma)}{(1+a\lambda\dot\gamma)(1-2a\lambda\dot\gamma)}\right)$	$4\eta\left(1+\frac{4a^2(\lambda-\Lambda)\lambda\dot\varepsilon^2}{1-4a^2\lambda^2\dot\varepsilon^2}\right)$	$2\eta\left(1-\frac{2a(\lambda-\Lambda)\dot\varepsilon}{1+2a\lambda\dot\varepsilon}\right)$

and elongation for some rate equation models.

and in strip biaxial extension the identical function

$$\eta_{sb} = \frac{G\lambda B}{2\lambda^2\dot\varepsilon^2} + O(\dot\varepsilon^{-4})$$

is obtained, while $\eta^x_{sb} \sim \dot\varepsilon^{-1}$. This has the unfortunate (or unlikely) property that the applied tensile stress decreases with increasing rate of extension in both uniaxial and strip biaxial elongation. The same ($\eta_T \sim \dot\gamma^{-2}$) is true of the Meister model but not of the Bogue-White model where $\eta_T \sim \dot\gamma^{-1}$, corresponding to the steady stress tending to a constant value as $\dot\gamma$ increases.

Model	*Defining equation*	*Number of constants*	$\Psi_1(\kappa)$	$\Psi_2(\kappa)$
Ward-Jenkins	(2.29)	3	$2G\lambda^2$	$\varepsilon G\lambda^2$
Bird-Carreau	(2.34)	3	$\dfrac{2G\lambda^2}{1+\tau^2\kappa^2}$	0
Carreau B	(2.35)	5	$2G\lambda^2 F(\kappa^2)g(\kappa^2)$	0
Meister	(2.32)	3	$\dfrac{2G\theta^2}{(1+b\theta\|\kappa\|)^3}$	0
Bogue-White	(2.36)	3	$\dfrac{2G\lambda^2}{(1+\frac{1}{2}a\lambda\|\kappa\|)^2}$	0
Tanner-Simmons	(2.37)	3	$2G\lambda^2\left[1-e^{-B/\lambda\kappa}\left(1+\dfrac{B}{\lambda\kappa}+\dfrac{B^2}{2\lambda^2\kappa^2}\right)\right]$	0

$$S = \alpha_2/\alpha_1 ,$$

Table 2.3 Material functions in steady shearing

For the Carreau model B the value S = 1 would give $\eta_T \sim \dot{\gamma}^{-2}$, and we need $S > \frac{1}{2} + R$ to allow the possibility of steady elongation at high rates, which gives $\eta_T \sim \dot{\gamma}^{-(1+2R)}$ and hence again for $R > 0$ a stress which decreases as $\dot{\gamma}$ increases.

In this discussion it should be remembered that we are looking at limiting cases which may be difficult or impossible to realise in practice, and behaviour of the sort we have found here and have claimed to be contrary to physical intuition does not provide very strong evidence for discarding or even mistrusting a model. Indeed the result that the stress begins to decrease with increasing strain rate at some strain rate could be taken as an indication of the beginning of instability or of necking and possibly failure of a material under tension. What the results do tell us is that we may expect

$\eta(\kappa)$	$\eta_T(\dot\gamma)$
$G\lambda$	$\dfrac{3G\lambda}{(1-2\lambda\dot\gamma)(1+\lambda\dot\gamma)}\left(1+\dfrac{\varepsilon\lambda\dot\gamma}{(1+2\lambda\dot\gamma)(1-\lambda\dot\gamma)}\right)$
$\dfrac{G\lambda}{1+\tau^2\kappa^2}$	$\dfrac{3G\lambda}{(1+3\tau^2\dot\gamma^2)(1-2\lambda\dot\gamma)(1+\lambda\dot\gamma)}$
$G\lambda F(\kappa^2)$	$\dfrac{3G\lambda F(3\dot\gamma^2)}{(1-2\lambda\dot\gamma g(3\dot\gamma^2))(1+\lambda\dot\gamma g(3\dot\gamma^2))}$
$\dfrac{G\theta}{(1+b\theta\lvert\kappa\rvert)^2}$	$\dfrac{3G\theta}{(1+\theta(b\sqrt{3}\lvert\dot\gamma\rvert-2\dot\gamma))(1+\theta(b\sqrt{3}\lvert\dot\gamma\rvert+\dot\gamma))}$
$\dfrac{G\lambda}{1+\frac{1}{2}a\lambda\lvert\kappa\rvert}$	$\dfrac{3G\lambda(1+\frac{1}{2}a\sqrt{3}\lambda\lvert\dot\gamma\rvert)}{(1+\lambda(\frac{1}{2}a\sqrt{3}\lvert\dot\gamma\rvert-2\dot\gamma))(1+\lambda(\frac{1}{2}a\sqrt{3}\lvert\dot\gamma\rvert+\dot\gamma))}$
$G\lambda\left[1-e^{-B/\lambda\kappa}(1+B/\lambda\kappa)\right]$	$\dfrac{G[3\lambda-\{B_T^2(1+\lambda\dot\gamma)-B_T^{-1}(1-2\lambda\dot\gamma)\}B_T^{-1/\lambda\dot\gamma}/\dot\gamma]}{(1-2\lambda\dot\gamma)(1+\lambda\dot\gamma)}$

$$F(\kappa^2) = \frac{1}{1+(\tau^2\kappa^2)^S}\ ,\quad g(\kappa^2) = \frac{(1+\lambda^2\kappa^2)^R}{1+(\tau^2\kappa^2)^S}\ ,\quad B_T^2+2B_T^{-1} = B^2+3\ ,$$

and elongation for some integral equation models.

similar results in analysing more complex flows and should be warned to exercise care in interpreting any results at high elongation rates for models with this behaviour.

2.2.3 Elongational viscosity and the existence of steady flow

We see that for many Maxwell and Jeffreys models the expression for the elongational viscosity has a denominator which may become zero (and then negative) as $\dot\gamma$ increases. A little care is needed in interpreting this. We may talk about an "infinite elongational viscosity" and a critical or "limiting strain rate" but only if we remember that we obtained these expressions by assuming that a steady flow existed, with constant stress and constant rate of strain. Study of the equations for time-dependent elongation (section 2.3,

Model	$\eta_{sb}(\dot{\varepsilon})$	$\eta^{x}_{sb}(\dot{\varepsilon})$
Ward-Jenkins	$\dfrac{4G\lambda}{1-4\lambda^2\dot{\varepsilon}^2}$	$\dfrac{2G\lambda}{(1+2\lambda\dot{\varepsilon})}\left(1-\dfrac{2\varepsilon\lambda\dot{\varepsilon}}{1-2\lambda\dot{\varepsilon}}\right)$
Bird-Carreau	$\dfrac{4G\lambda}{(1+4\tau^2\dot{\varepsilon}^2)(1-4\lambda^2\dot{\varepsilon}^2)}$	$\dfrac{2G\lambda}{(1+4\tau^2\dot{\varepsilon}^2)(1+2\lambda\dot{\varepsilon})}$
Carreau B	$\dfrac{4G\lambda F(4\dot{\varepsilon}^2)}{1-4\lambda^2\dot{\varepsilon}^2 g(4\dot{\varepsilon}^2)^2}$	$\dfrac{2G\lambda F(4\dot{\varepsilon}^2)}{1+2\lambda\dot{\varepsilon}g(4\dot{\varepsilon}^2)}$
Meister	$\dfrac{4G\theta}{1+4b\theta\lvert\dot{\varepsilon}\rvert+4(b^2-1)\theta^2\dot{\varepsilon}^2}$	$\dfrac{2G\theta}{(1+2\theta(b\lvert\dot{\varepsilon}\rvert+\dot{\varepsilon}))(1+2b\theta\lvert\dot{\varepsilon}\rvert)}$
Bogue-White	$\dfrac{4G\lambda(1+a\lambda\lvert\dot{\varepsilon}\rvert)}{(1+\lambda(a\lvert\dot{\varepsilon}\rvert+2\dot{\varepsilon}))(1+\lambda(a\lvert\dot{\varepsilon}\rvert-2\dot{\varepsilon}))}$	$\dfrac{2G\lambda}{1+\lambda(a\lvert\dot{\varepsilon}\rvert+2\dot{\varepsilon})}$
Tanner-Simmons	$\dfrac{G[4\lambda-\{B_{sb}^{2}(1+2\lambda\dot{\varepsilon})-B_{sb}^{-2}(1-2\lambda\dot{\varepsilon})\}B_{sb}^{-1/\lambda\dot{\varepsilon}}/\dot{\varepsilon}]}{1-4\lambda^2\dot{\varepsilon}^2}$	$\dfrac{G[2\lambda+B_{sb}^{(-2-1/\lambda\dot{\varepsilon})}/\dot{\varepsilon}]}{1+2\lambda\dot{\varepsilon}}$

$$B_{sb}^{2}+B_{sb}^{-2} = B^{2}+2 .$$

Table 2.3 (continued)

below) shows us that in fact for a strain rate in excess of the critical value elongational flow with that strain rate may only be maintained by applying an exponentially increasing stress (see (2.119) and also [D10],[P16]). The corresponding integral equations (e.g. (2.26a) with (2.27b)) lead to the same conclusion, and if we assume again that we have steady elongation which has been maintained for an indefinite time we find the integrals do not converge, as noted by Lodge ([L15],p. 116). Similar remarks are true for strip-biaxial extension (pure shear) at least as far as η_{sb} is concerned, though the "cross-viscosity" and the transverse force do not become infinite for upper convected models or for generalised models with $0 \leq a \leq 1$. (We consider $\dot{\varepsilon} > 0$ only.) Likewise in equal biaxial extension by considering η_T for negative values of $\dot{\gamma}$ we see the same behaviour (with different critical strain rates for η_{eb}.)

Equation (2.52), for the 8-constant Oldroyd model does have a denominator which need not necessarily be zero. If

$$(3\nu_1 - 2\mu_1)(\mu_1 - 3\mu_o/2) > \mu_1^2/4 \tag{2.54}$$

then the quadratic in $\dot{\gamma}$ (which is the denominator of (2.52)) has no real roots. However inequality (2.54) is not sufficient to ensure the existence of steady elongation at any strain rate, as Pountney has shown [P30] for the 4-constant model (with $\nu_1 = \nu_2 = 0$, $\mu_1 = \lambda_1$, $\mu_2 = \lambda_2$). In this case the inequality becomes

$$4(3\mu_o - 2\lambda_1)\lambda_1 > \lambda_1^2 \tag{2.55a}$$

or

$$\mu_o > 3\lambda_1/4 \ . \tag{2.55b}$$

Examination of the behaviour of the time-dependent part of the stress in the start-up of uniaxial elongation at constant rate of strain reveals that, if

$$\underline{p} = \underline{P}\exp(-\alpha t/\lambda_1) \tag{2.56}$$

then α satisfies the characteristic equation

$$\alpha^2 - (2 - \lambda_1\dot{\gamma})\alpha + 1 - \lambda_1\dot{\gamma} + \lambda_1(3\mu_o - 2\lambda_1)\dot{\gamma}^2 = 0 \tag{2.57}$$

Now clearly the question of whether $\underline{p}(t)$ approaches a steady value depends on whether the roots of (2.57) are positive. When $\lambda_1\dot{\gamma} > 2$ the sum of the roots is negative, so this puts an upper limit on the strain rate for which a steady state may be attained. Figure 2.2 shows the values of μ_o/λ_1 and $\lambda_1\dot{\gamma}$ for which steady flow is possible, with negative values of $\dot{\gamma}$ corresponding to equal biaxial extension.

Of the models in Table 2.3, there is a "critical elongation rate" and an "infinite elongational viscosity" always for the Ward-Jenkins and Bird-Carreau models, sometimes (depending on the parameters) for the Carreau model B, Meister and Bogue-White models and never (except in degenerate cases) for the Tanner-Simmons model. The conditions for a finite elongational viscosity are $b > 2/\sqrt{3}$ for the Meister model, $a > 4/\sqrt{3}$ for the Bogue-White model and $2\lambda\dot{\gamma}\, g(3\dot{\gamma}^2) < 1$ (for all $\dot{\gamma}$) for the Carreau model B.

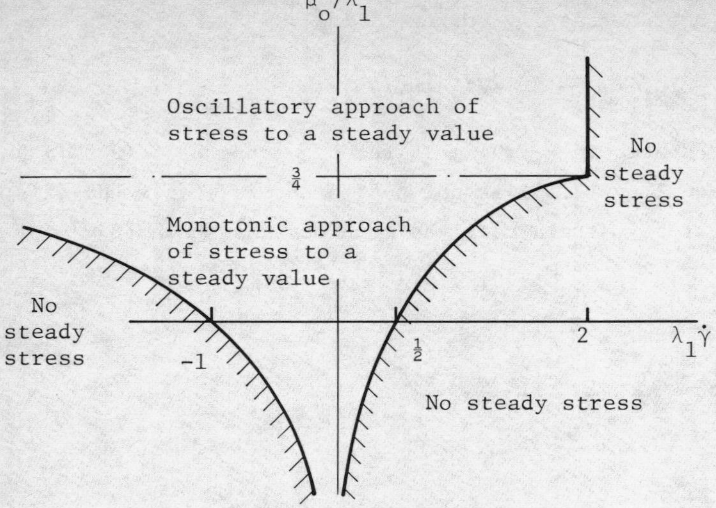

Figure 2.2 Stability diagram for an Oldroyd model.

Note however that the full Kaye-BKZ model with the function (3.60c) of Zapas does contain terms which give an elongational viscosity which becomes infinite (at strain rates $1/2\lambda$ for the terms with c' and $1/4\lambda$ for the term with α'). This is contrary to the statements of several authors [T8, D3, C3] who appear to be considering the modified Kaye-BKZ equation of Adams and Bogue [A10] which has $\alpha' = c' = 0$. We have not tabulated the material functions for the Kaye-BKZ model (2.38) because they involve complicated integrals arising from the logarithmic terms in the equation (2.38c) for U, and the other terms give expressions the same as and similar to those we have for the convected Maxwell models.

We emphasise that the elongational viscosity as defined here is a material property, both for our model liquids and for real liquids. It must be admitted that in practice its measurement may be extremely difficult because it requires both a constant rate of strain and a constant stress. Furthermore, as we have seen for the Oldroyd model, the existence of a well behaved function purporting to be an elongational viscosity does not guarantee that we can obtain the required steady flow in an experiment. The converse, that if the elongational viscosity tends to infinity as the strain rate approaches a critical value then strain rates above the critical value cannot be maintained with a constant applied stress, is not proved, but is true for all cases investigated.

2.2.4 Further constitutive equations

There are many recent and some not so recent constitutive equations which we have omitted from the discussion in this chapter, notably a number of integral equations similar to the Kaye-BKZ equation (2.38). These equations, which can be thought of either in terms of a strain-dependent memory function, or if the strain and time-dependence of the memory function may be separated we may think of them as models with a fixed memory function and a nonlinear deformation tensor. This is discussed by the author in a recent paper [P20] and many papers are cited there, including [D23,P24,W 2 ,Y 3]. Elongational flow properties are discussed [D23,P24,W 2] but explicit formulae are not obtained. There is not space to give these papers the discussion they deserve. We remark only that Doi and Edwards [D23] use a molecular model which gives the most detailed consideration to date of the process of entanglement formation and breakage in a network of entangled molecules. The work of Yamamoto [Y1,Y2] merits further study, since this early network theory is sufficiently general to allow any of the forms of strain rate dependence of elongational viscosity reported here.

In this section we discuss some Maxwell models, with the particular aim of seeing how the upper limit to shear rates attainable in steady flow is affected by variable parameters. The commonest approach is that associated with White and Metzner [W10] who proposed the equation

$$\underline{p} + \lambda(II_e)\overset{\nabla}{\underline{p}} = 2G\lambda(II_e)\underline{e} \tag{2.58}$$

with constant modulus G and relaxation time λ dependent on the invariant II_e of the rate of strain tensor. Tanner [T 6] proposed a similar model with λ constant and G variable, and more recently Metzner and co-workers [A11,B 1] have used independent empirically determined functions $\lambda(II_e)$ and $\eta(II_e)$ (in place of $G\lambda(II_e)$ in (2.58)). In Table 2.4 we show three functions which have been used either in equations like (2.58) or in similar integral equations. The function $f(\varepsilon)$ introduced by Baid and Metzner is zero in shear and positive in elongation, but no details are given. For all the experimental results [B 1] a value $\theta_4 f(\varepsilon) = 0.07$ seconds was found to be adequate.

Reference	Spearot [S15] cf Bird-Carreau [B16]	Baid [B1]	cf Bogue [C13], Meister [M25]
Relaxation time, λ	$\frac{\theta}{1+CII_e}$	$\frac{\theta_3}{1+C\theta_3^2 II_e} + \theta_4 f(\varepsilon)$	$\frac{\theta}{1+2b\theta\sqrt{II_e}}$
Viscosity	$G\lambda(II_e)$	$\eta_o\left[\frac{1+\theta_2^2 II_e}{1+\theta_1^2 II_e}\right]$	$G\lambda(II_e)$
η_T (small $\dot{\gamma}$)	$3G\theta\{1+\theta\dot{\gamma}+3(\theta^2-C/4)\dot{\gamma}^2\}$	$3\eta_o\{1+(\theta_3+\theta_4 f(\varepsilon))\dot{\gamma}\}$	$3G\theta\{1+\theta(\dot{\gamma}-b\sqrt{3}\lvert\dot{\gamma}\rvert)\}$
critical $\dot{\gamma}$ ($\eta_T\to\infty$)	None for $C > 4\theta^2/3$	If $\theta_4 f(\varepsilon)$ is constant there must be a critical $\dot{\gamma}$	None for $b > 2/\sqrt{3}$
η_T ($\dot{\gamma}\to+\infty$)	$4G\ /C\dot{\gamma}^2$	-	$\frac{3Gb\sqrt{3}}{(b\sqrt{3}-2)(b\sqrt{3}+1)\theta\dot{\gamma}}$
η_{sb} (small $\dot{\varepsilon}$)	$4G\theta\{1+4\theta^2\dot{\varepsilon}^2\}$	$4\eta_o\{1+4(\theta_3+\theta_4 f(\varepsilon))^2\dot{\varepsilon}^2\}$	$4G\theta\{1+4\theta^2\dot{\varepsilon}^2\}$
η_{sb}^x (small $\dot{\varepsilon}$)	$2G\theta\{1-2\theta\dot{\varepsilon}+4\theta^2\dot{\varepsilon}^2\}$	$2\eta_o\{1-2(\theta_3+\theta_4 f(\varepsilon))^2\dot{\varepsilon}\}$	$2G\theta\{1-2\theta\dot{\varepsilon}+4(b+1)\theta^2\dot{\varepsilon}^2\}$
η_{sb} ($\dot{\varepsilon}\to+\infty$)	$\frac{4G\theta}{C\dot{\varepsilon}^2}$	-	$\frac{2G}{(b^2-1)\dot{\varepsilon}}$
η_{sb}^x ($\dot{\varepsilon}\to+\infty$)	$\frac{2G\theta}{C\dot{\varepsilon}^2}$	-	$\frac{G}{(b+1)\dot{\varepsilon}}$

Table 2.4 Elongational viscosity for Maxwell models with strain-rate-dependent parameters.

In fact the original suggestion of White and Metzner [W10] was for a stress-dependent relaxation time and, while this was ignored for some time because it is more difficult to handle equations which only give the extra-stress implicitly, recent work has returned to this theme. Table 2.5 shows some elongational viscosity results for three models. The models of Phan-Thien and Tanner [P21] and Phan-Thien [P22] are both of the form of (2.58) with

Reference	Phan-Thien [P21]	Phan-Thien [P22]	Marrucci [A4]
Relaxation time, λ	$\dfrac{\lambda_o}{1+bI_p}$	$\lambda_o \exp(-bI_p)$	$\lambda_o x^{1.4}$
Viscosity	$\eta_o \lambda(I_p)/\lambda_o$	$\eta_o \lambda(I_p)/\lambda_o$	$\eta_o x^{2.4}$
Small strain rate approximations ($\dot\gamma > 0,\ \dot\varepsilon > 0$)			
$\lambda(\dot\gamma)$	$\lambda_o(1-6ab\eta_o\lambda_o\dot\gamma^2)$		$\lambda_o(1-1.4a\sqrt{3}\lambda_o\dot\gamma)$
$\eta_T(\dot\gamma)$	$3\eta_o(1+a\lambda_o\dot\gamma+3a^2\lambda_o^2\dot\gamma^2)$		$3\eta_o(1+(1-2.4a\sqrt{3})\lambda_o\dot\gamma)$
$\lambda(\dot\varepsilon)$	$\lambda_o(1-8ab\eta_o\lambda_o\dot\varepsilon^2)$		$\lambda_o(1-2.8a\lambda_o\dot\varepsilon)$
$\eta_{sb}(\dot\varepsilon)$	$4\eta_o(1+4a^2\lambda_o^2\dot\varepsilon^2)$		$4\eta_o(1-4.8a\lambda_o\dot\varepsilon)$
$\eta^x_{sb}(\dot\varepsilon)$	$2\eta_o(1-2a\lambda_o\dot\varepsilon+4a^2\lambda_o^2\dot\varepsilon^2)$		$2\eta_o(1-(2+4.8a)\lambda_o\dot\varepsilon)$
Large strain rate approximations ($\dot\gamma > 0,\ \dot\varepsilon > 0$)			
$\lambda(\dot\gamma)$*	$\dfrac{1}{2a\dot\gamma}\left[1-\dfrac{b\eta_o}{2a^2\lambda_o^2\dot\gamma^2}\right]$	$\dfrac{1}{2a\dot\gamma}\left[1-\dfrac{b\eta_o}{a\lambda_o\log(2a\lambda_o\dot\gamma)}\right]$	$\dfrac{1}{2\dot\gamma}\left[1-\dfrac{a^2}{2}(2\lambda_o\dot\gamma)^{-2/1.4}\right]$
$\eta_T(\dot\gamma)$*	$\dfrac{2\lambda_o}{b}$	$\dfrac{\log(2a\lambda_o\dot\gamma)}{b\dot\gamma}$	$\dfrac{4\eta_o}{a^2}(2\lambda_o\dot\gamma)^{-0.4/1.4}$
$\eta^x_{sb}(\dot\varepsilon)$	$\dfrac{\eta_o}{2a\lambda_o\dot\varepsilon}$	$\dfrac{\eta_o}{2a\lambda_o\dot\varepsilon}$	$\eta_o(2\lambda_o\dot\varepsilon)^{-2.4/1.4}$

* In strip biaxial extension at large strain rates $\lambda(\dot\varepsilon) = \lambda(\dot\gamma)$ and $\eta_{sb}(\dot\varepsilon) = \eta_T(\dot\gamma)$ if $\dot\varepsilon = \dot\gamma$.

Table 2.5 Elongational viscosity for Maxwell models with stress-dependent parameters.

$\lambda(II_e)$ replaced by $\lambda(I_p)$ and with the upper convected derivative replaced by the generalised convected derivative used in (2.21). (Each of the actual models in Table 2.5 has a spectrum of relaxation times, but we continue to compare results for a single mode for simplicity.) The model of Marrucci and his co-workers [M12,A4,A5,A7] has a structure parameter x in it, and the parameter a there is not associated with the convected derivative but with the stress-dependent breakdown of structure. For one relaxation time the model may be written

$$\underline{p}/G + \lambda \overset{\nabla}{\overline{(\underline{p}/G)}} = 2\lambda \underline{e} , \tag{2.59a}$$

$$\lambda = \lambda_o x^{1.4} , \quad G = G_o x , \tag{2.59b,c}$$

$$\lambda \dot{x} = 1 - x - ax \sqrt{[I_p/2G]} . \tag{2.59d}$$

and in Table 2.5 we write η_o for $\lambda_o G_o$. The invariant I_p ($I_p \equiv tr(\underline{p})$) may be associated with the stored elastic energy (entropic free energy) in a network of entangled molecules.

If we consider flow at constant strain-rate and ask whether a constant stress is reached, the criterion for the models in Table 3.4 is the same as for models with constant spectra, and the conditions

$$1 - 2\lambda(II_e)\dot{\gamma} > 0 , \quad 1 + \lambda(II_e)\dot{\gamma} > 0 \tag{2.60a,b}$$

are necessary and sufficient for the existence of steady flow with a constant stress and for the existence of a finite positive elongational viscosity. For the Phan-Thien models and the Marrucci model in Table 2.5 the dependence of the relaxation time on the stress makes the start-up of flow at a given strain rate more difficult to analyse and we cannot make as strong a statement. Computation by these authors [P21,P22,A4,A5] and some analysis does appear to show that for these models the elongational viscosity is always bounded and that a steady stress is obtained.

In obtaining the approximate results in Table 2.5 for the stress dependent models we use the steady state values (for the generalized convected Maxwell model (2.21))

$$I_p \equiv tr(\underline{p}) = 6aG\lambda^2\dot{\gamma}^2/(1-2a\lambda\dot{\gamma})(1+a\lambda\dot{\gamma}) \tag{2.61}$$

in uniaxial elongation, and

$$I_p \equiv tr(\underline{p}) = 8aG\lambda^2\dot{\varepsilon}^2/(1-4a^2\lambda^2\dot{\varepsilon}^2) \tag{2.62}$$

in strip biaxial elongation. For the Marrucci model we set a = 1; the parameter a in (2.59) is different. Then the structure parameter x is obtained from the steady-state solution of (2.59d), namely from

$$(1 - x)^2 = a^2x^2I_p/2G \tag{2.63}$$

using $\lambda = \lambda_o x^{1.4}$. Numerical solutions of the nonlinear algebraic equations, in the original papers, give results consistent with the approximate results obtained here by direct substitution (of undetermined functions).

We remark on two aspects of the results for the variable parameter Maxwell models. In all cases the relaxation time decreases with increasing rate of strain. If this decrease is rapid enough the strain rate will never reach the critical value of $1/2a\lambda$. The Baid expression for the relaxation time has a lower bound (if we take $\theta_4 f(\varepsilon)$ to be constant as he does [B 1]) and hence eventually $\dot{\gamma}$ will reach $1/2\lambda$. To avoid this it would be necessary to allow $\theta_4 f(\varepsilon)$ to become as small as necessary for large strain rate. Also note that the applied stress is $\eta_T\dot{\gamma}$, so that the Spearot form of λ gives us a decreasing stress with increasing $\dot{\gamma}$ (for large $\dot{\gamma}$). We have discussed this problem above (section 2.2.2) in discussing integral equation models. The stress-dependent models all have elongational stress increasing with strain rate. The fact that the transverse stress in strip biaxial extension tends to a constant value is not physically unreasonable, nor indeed is the decrease of this stress with increasing strain rate which we see with the Marrucci model. (At least it is not obviously contrary to any physical law.)

2.3 BASIC ELONGATIONAL FLOWS

In this section we shall analyse the three basic elongational flows, uniaxial stretching, biaxial stretching and spinning. In order to provide a useful framework for the discussion of experimental results in chapter 3 below we use three very simple constitutive equations, the Newtonian liquid, the upper convected Maxwell liquid and the co-rotational Maxwell liquid. The basic tactic we adopt here is to postulate a plausible motion in terms of one or two kinematic variables which depend on one independent variable (time or one spatial coordinate). Then substitution of this into the equations of conservation of momentum and the constitutive equation gives us a system of ordinary differential equations to solve, subject to suitable boundary and initial conditions. We shall generally neglect the effects of inertia, gravity and surface tension. More complex problems are discussed below in chapter 6 and to some extent in chapter 5. The continuity equation (conservation of mass) is not usually needed explicitly because we choose motions which satisfy it identically.

2.3.1 Uniaxial stretching

We consider here a flow which is time-dependent but spatially homogeneous, in the sense that the rate of strain tensor $\underline{e}$ is independent of position $\underline{X}$. This tensor is a time-dependent version of (2.42a), namely

$$\underline{e}(t) = \begin{pmatrix} \dot{\gamma}(t) & 0 & 0 \\ 0 & -\frac{1}{2}\dot{\gamma}(t) & 0 \\ 0 & 0 & -\frac{1}{2}\dot{\gamma}(t) \end{pmatrix} \qquad (2.64)$$

where $\dot{\gamma}(t)$ is the rate of strain, and if we consider stretching a cylindrical

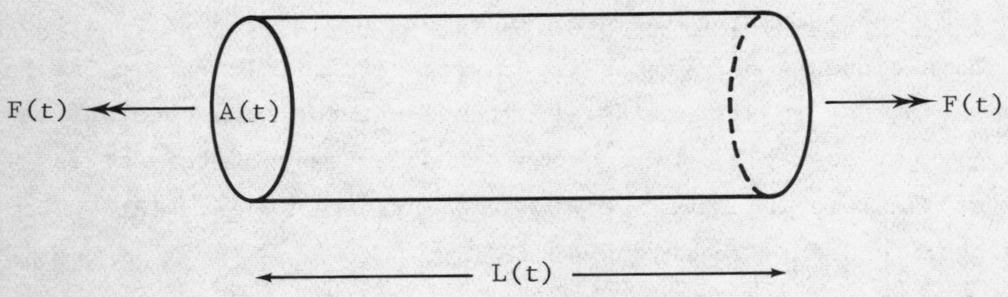

Figure 2.3 Uniaxial stretching.

specimen of length L and cross-sectional area A (Figure 2.3) then

$$\dot{\gamma} = \frac{1}{L}\frac{dL}{dt} = -\frac{1}{A}\frac{dA}{dt} \qquad (2.65)$$

Note that the velocity at any point is a linear function of $\underline{X}$ and not a constant. However since rate of strain and stress tensors are independent of $\underline{X}$ we have $dp_{ii}/dt = \partial p_{ii}/\partial t$ and the velocity does not appear explicitly in the equations when inertia may be neglected.

There are a variety of modes of stretching which we shall consider; constant rate-of-strain, $\dot{\gamma}$, constant velocity of the end of the specimen, dL/dt, constant stress,

$$T \equiv p_{11} - p = F/A \ , \tag{2.66}$$

(see section 2.3.4 below) or constant force, F. It is important (unless strains are very small) to distinguish clearly between constant rate of strain and constant velocity and between constant stress (F/A) and constant force, sometimes referred to as constant engineering stress (F/A_o) where A_o is the initial value of A. In solid mechanics, including some work on polymers, this confusion is possible and careful reading of descriptions is essential. It is worth noting, for example, that $L_o^{-1}(dL/dt)$ is referred to as the rate of strain in some publications, where the initial length L_o replaces the time dependent length L of our (correct) definition.

2.3.2 Biaxial stretching

Under this heading we could consider quite general stretching flows, but we shall restrict ourselves to two special cases, namely equal biaxial stretching and strip biaxial extension or pure shear. The former is kinematically the reverse of uniaxial stretching, but dynamically can be quite different. The rate of strain tensor is, like (2.43a),

$$\underline{e}(t) = \begin{pmatrix} -2\dot{\Gamma}(t) & 0 & 0 \\ 0 & \dot{\Gamma}(t) & 0 \\ 0 & 0 & \dot{\Gamma}(t) \end{pmatrix} \tag{2.67}$$

where the equal stretching in the X_2 and X_3 directions is at a strain rate $\dot{\Gamma}(t)$ which is positive. This is the same as (2.64) for uniaxial stretching if we set $\dot{\gamma} = -2\dot{\Gamma} < 0$. As we shall see, the behaviour of the stress tensor as a function of $\dot{\gamma}$ is different for negative $\dot{\gamma}$ for the upper convected Maxwell model.

The applied stress, T, is equal in the X_2 and X_3 directions (from the geometrical symmetry of the situation) so that we have

$$T(t) = p_{22} - p = p_{33} - p \tag{2.68}$$

For a square sheet of side L(t) and thickness H(t) (Figure 2.4) we have an applied force

$$F(t) = T(t)\ L(t)\ H(t) \tag{2.69}$$

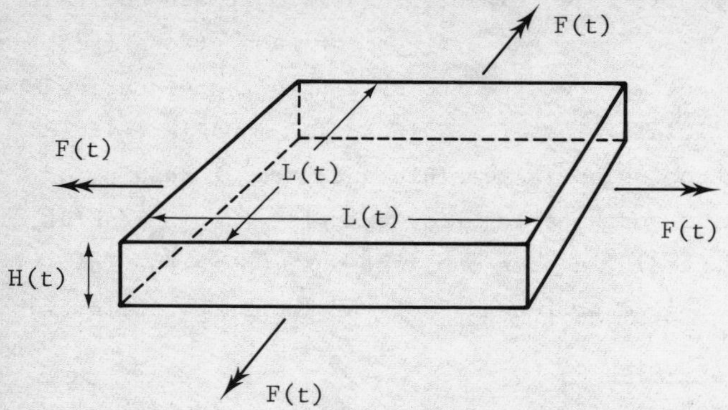

Figure 2.4 Equal biaxial stretching.

in each of the X_2 and X_3 directions; mass conservation gives

$$L^2(t)\ H(t) \ = \ V_M \tag{2.70}$$

where V_M is the constant volume of material; and

$$\dot{\Gamma} = \frac{1}{L}\frac{dL}{dt} \ = \ -\frac{1}{2H}\frac{dH}{dt} \ . \tag{2.71}$$

The other kinematics we consider are those of strip biaxial extension, also known as pure shear, which is an irrotational two-dimensional motion with the time-dependent version of (2.44a),

$$\underline{e}(t) \ = \ \begin{pmatrix} \dot{\varepsilon}(t) & 0 & 0 \\ 0 & -\dot{\varepsilon}(t) & 0 \\ 0 & 0 & 0 \end{pmatrix} \tag{2.72}$$

and applied stress $T = p_{11} - p$. This may be realised by stretching a strip in the X_1 direction and restraining it in the X_3 direction or by the four-roll mill of G. I. Taylor [T15]. The idea of the latter is most easily explained in terms of streamlines, so we write down the velocity field corresponding to (2.32);

$$U_1 = \dot{\varepsilon}X_1 , \quad U_2 = -\dot{\varepsilon}X_2 . \tag{2.73}$$

The streamline at (X_1,X_2) has slope

$$\frac{dX_2}{dX_1} = \frac{-X_2}{X_1}$$

which has the obvious solution $X_1 X_2$ = constant; in other words the streamlines are the family of rectangular hyperbolae with the axes as asymptotes. If four cylinders are immersed in a liquid with axes through the four

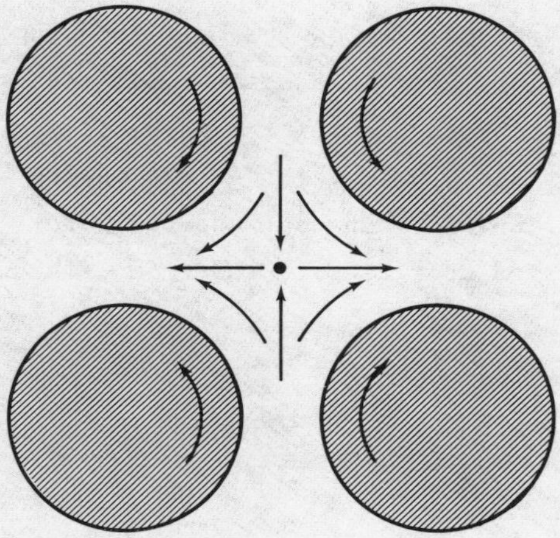

Figure 2.5 *The four-roll mill; pure shear or planar extension.*

corners of a square and each is rotated in the direction opposite to the two adjacent ones, the velocity field between them will be of the required form with a stagnation point at the centre, as illustrated in Figure 2.5. If the cylinders are long enough the flow will be essentially two-dimensional. This flow clearly is attractive to the rheologist since it can be realised with less viscous materials than is the case with filament and sheet extension. However we shall run into problems because the velocity gradient cannot be uniform right up to the cylinder boundaries - we have to match hyperbolic streamlines and circular boundaries which are also streamlines.

We shall not therefore discuss this further but refer the reader to the few published accounts of work with this and related geometries [T15,G7,G9,M27, B26,B12,M7,P29].

2.3.3 Spinning

We continue to make the basic assumptions that inertia, gravity and surface tension may be ignored, and that the stress and axial velocity are uniform across a cross-section. We are now considering a flow which is steady (in a laboratory frame of reference) in which a jet or thread of liquid emerges from an orifice into air (or some fluid much less viscous than the liquid of the jet) and is stretched by a force applied in some way at a fixed point some distance downstream of the orifice. As an element of fluid in the jet flows from the orifice to the take-up point (as we term the point where the force is applied, using the terminology of melt spinning) it is stretched under the action of the force, its cross-sectional area, A, decreases and its velocity, U, increases, though all these quantities at any point in space are independent of time. There is a close analogy with uniaxial stretching under the action of a constant applied force (and hence a stress which increases as the area decreases). The flow is shown schematically in Figure 2.6.

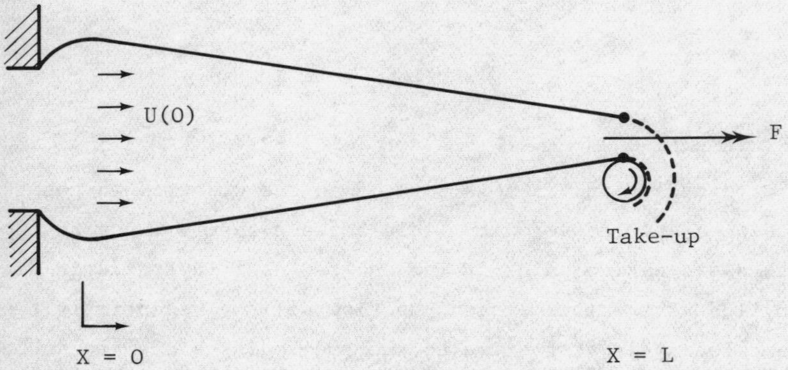

Figure 2.6 Sketch of the spinning process (schematic).

For convenience we drop the suffix 1 on X_1 and the corresponding velocity U_1 and denote dU/dX by U'(X). U, A, p_{ii} and e_{ii} are functions of X only and the material derivative is given by $dp_{ii}/dt = U\, dp_{ii}/dX$. The components of

$$\underline{e}(X) = \begin{pmatrix} U'(X) & 0 & 0 \\ 0 & -\tfrac{1}{2}U'(X) & 0 \\ 0 & 0 & -\tfrac{1}{2}U'(X) \end{pmatrix} \qquad (2.74)$$

similar to (2.42a) and (2.64). We need to choose appropriate boundary conditions to complete the problem specification, and this does present some real difficulties. We should like to specify the initial value U(0) at the orifice where we know A(0), the orifice cross-sectional area, and hence might expect U(0) = Q/A(0) to be reasonable with volume flow rate Q prescribed independently of the rest of the experiment. This is not possible because at the orifice our assumption of a velocity uniform across the cross-section cannot hold. The transition from shearing flow to elongational flow at the exit from a tube or orifice is a well-known unsolved problem even for a Newtonian liquid and the problems are apparently increased for elasticoviscous liquids with the phenomenon of extrudate swell and clear indications that such things as the length of the tube can have a large effect on the initial diameter of the jet.

We shall avoid this problem by taking an origin for X some distance downstream of the orifice as indicated in Figure 2.6, and measure A(0) and U(0) directly or indirectly. This leaves us vulnerable to criticism when we discuss changes in applied force and assume that A(0) and U(0) are unaffected, but it is the best we can do at present. For the Maxwell models (and other elasticoviscous liquids) there will be other initial conditions required, and we shall be even more uncertain about these since they may involve quantities we cannot measure directly. For the moment we shall just prescribe what we need, and postpone discussion to chapter 6.

Another question to be postponed concerns the force F. In the simple case (no gravity or inertia) we can in principle prescribe it, by applying a force F at the take-up. If inertia or gravity is important the force on a cross-section varies down the jet and we can only prescribe it at the take-up, giving us in effect a two-point boundary value problem. Even in the simple case, though, it is more convenient in most experiments to prescribe the take-up velocity, and then the force cannot be prescribed independently, but is determined by the take-up velocity and other parameters of the flow. This will become clear in the analysis below.

2.3.4 Modelling the basic flows

Since the rate of strain tensor $\underline{e}$ is diagonal and the vorticity tensor is zero, we may show that the extra stress tensor $\underline{p}$ is diagonal and the constitutive equations simplify to

$$p_{ii} = 2\eta e_{ii} \tag{2.75}$$

$$p_{ii} + \lambda\{dp_{ii}/dt - 2e_{ii}p_{ii}\} = 2\eta e_{ii} \tag{2.76}$$

$$p_{ii} + \lambda dp_{ii}/dt = 2\eta e_{ii} \tag{2.77}$$

for the Newtonian, upper convected Maxwell and co-rotational Maxwell liquids respectively. Here there is no summation on the repeated suffix i, dp_{ii}/dt is the material derivative (following a material particle), η is the viscosity of the liquid and λ its relaxation time. The material parameters η and λ are taken to be constants, and since we assume isothermal flow we ignore the energy equation.

In the cases of uniaxial stretching and spinning we assume that stresses do not vary across a cross-section, and then the local equation of motion in the direction of elongation may be integrated across the cross-section to give the condition that the force on a cross-section is independent of distance in the direction of flow, $X \equiv X_1$,

$$(p_{11} - p)A \equiv TA = F \tag{2.66}$$

where A is the cross-sectional area (a function of X or time t), F the force (a function of t only) and p is the isotropic pressure (undetermined by the constitutive equation since the material is incompressible). This isotropic pressure is determined by using the condition of continuity of stress at the free surface,

$$p_{22} - p = p_{33} - p = 0 \tag{2.78}$$

where we are measuring pressures relative to atmospheric pressure, so the stress outside the free surface is zero. (The directions X_2 and X_3 are any two directions normal to X_1, and the choice is immaterial because we assume an axisymmetric flow.)

Hence we shall need two stress components from (2.75), (2.76) or (2.77) (for i = 1 and i = 2) using a chosen form of $\underline{e}$ with an unknown velocity function U, and these will be solved together with

$$(p_{11} - p_{22})A = TA = F . \tag{2.79}$$

Either

$$A(X)U(X) = Q \tag{2.80}$$

which is the integrated continuity equation in spinning, or

$$A(t)L(t) = V_M \tag{2.81}$$

which is the expression of the conservation of mass in stretching, already used to deduce (2.65), will be used to eliminate the variable A.

In equal biaxial stretching (2.66) is replaced by (2.68) and (2.78) by

$$p_{11} - p = 0 \tag{2.82}$$

since the free surface is normal to the X_1 axis, so (2.79) is replaced by

$$p_{22} - p_{11} = p_{33} - p_{11} = T(t) \tag{2.83}$$

with (2.69) and (2.70) used to relate T to the applied force and strain rate (2.71). Similarly in strip biaxial stretching the free surface is normal to the X_2 axis and

$$p_{22} - p = 0 \tag{2.84}$$

gives the isotropic pressure. The applied stresses are then

$$T(t) = p_{11} - p_{22} \tag{2.85}$$

in the X_1 direction and, preventing motion in the X_3 direction

$$T^X(t) = p_{33} - p_{22} \tag{2.86}$$

Notice how the principal stress differences used in the definition of the elongational viscosity, η_T, and biaxial elongational viscosities, η_{eb}, η_{sb}, η_{sb}^X, all are both quantities which are determined by the motion and the constitutive equation alone and quantities which may be measured in experimental realisations of the flows we have described (if steady conditions can be attained).

We can write down the equations for a general elongational flow, with rate of strain tensor

$$\underline{e} = \begin{pmatrix} e_1 & 0 & 0 \\ 0 & e_2 & 0 \\ 0 & 0 & e_3 \end{pmatrix} \tag{2.87}$$

Then the principal stress differences are given by

$$p_{ii} - p_{jj} = 2\eta(e_i - e_j) \tag{2.88}$$

for the Newtonian liquid, and

$$p_{ii} - p_{jj} + \lambda(\dot{p}_{ii} - \dot{p}_{jj}) = 2\eta(e_i - e_j) \tag{2.89}$$

for the co-rotational Maxwell model, where $\dot{p}_{ii}$ denotes dp_{ii}/dt in the stretching flows and Udp_{ii}/dX in spinning. For the upper convected model we need to perform some algebraic manipulation to get expressions involving only differences between the principal stresses, and we do this for each flow individually.

2.3.5 The Newtonian liquid

Use of (2.64 and (2.66) in (2.88) gives us for uniaxial stretching

$$T = p_{11} - p_{22} = 3\eta\dot{\gamma} \tag{2.90}$$

and hence from (2.79)

$$3\eta\dot{\gamma}A = F . \tag{2.91}$$

Finally (2.65) leads to

$$\frac{dA}{dt} = -\frac{F}{3\eta} . \tag{2.92}$$

Clearly if the applied stress, $T = F/A$, is constant then the rate of strain $\dot{\gamma}$ is constant and conversely if $\dot{\gamma}$ is constant the stress is constant, and their ratio is 3η, as proved by Trouton and tabulated above. The specimen length increases exponentially and the area decreases, keeping AL (the volume of material in the specimen) constant.

$$L = L_o e^{\dot{\gamma}t} = L_o e^{Tt/3\eta} . \tag{2.93}$$

If the force is constant then (2.92) is integrated to give

$$A = A_o - Ft/3\eta \tag{2.94}$$

and we see that in this case $A \to 0$, and $L \to \infty$, as $t \to 3\eta A_o/F$. Finally if the velocity $U = dL/dt$ (of the end of the specimen) is constant then the stress

$$T = 3\eta U/(L_o + Ut) \tag{2.95}$$

and the force

$$F = 3\eta U A_o L_o/(L_o + Ut)^2 \tag{2.96}$$

decreases monotonically with time. A special result for the Newtonian liquid, because it is purely viscous, is that the ratio $T(t)/\dot{\gamma}(t) = 3\eta$ always, whether the flow is steady or not. This is not the case for elasticoviscous liquids.

In spinning we get (2.90) again, with $\dot{\gamma}(t)$ replaced by $U'(X)$, and the use of (2.79) and (2.80) gives

$$3\eta U'(X)Q/U(X) = F \tag{2.97}$$

with the solution, satisfying the initial condition $U(0) = U_o$,

$$U(X) = U_o e^{FX/3\eta Q} \tag{2.98}$$

If, instead of a given force F, we have an additional boundary condition

$$U(L) = U_o D_R \tag{2.99}$$

where D_R is the given draw ratio $U(L)/U(0)$, we find F from (2.47)

$$F = \frac{3\eta Q}{L} \log D_R \tag{2.100}$$

Now for the Newtonian liquid we may evaluate the stress and the strain rate at any point, and find

$$T(X) = F/A(X) = (FU_o/Q)e^{FX/3\eta Q} \tag{2.101}$$

$$U'(X) = (FU_o/3\eta Q)e^{FX/3\eta Q} . \tag{2.102}$$

Hence we obtain again $T(X)/U'(X) = 3\eta$, the constant elongational viscosity of the Newtonian liquid.

It is possible to change variable from X to τ, where τ is the time that a particle of the liquid has been in the flow. We have

$$U = dX/d\tau$$

and hence

$$\tau(X) = \int_o^X U^{-1} dX$$

$$= (3\eta Q/FU_o)(1 - e^{-FX/3\eta Q}) \qquad (2.103a)$$

$$= (3\eta A_o/F)(1 - A(X)/A_o) \qquad (2.103b)$$

where $A_o = Q/U_o$. This can be rearranged to give equation (2.94) for constant force stretching (if we identify τ and t). We see then that the exponential decrease of area with distance corresponds to a linear decrease of area with time. Many such relations are given by Hyun and Ballman [H21].

For equal biaxial stretching we may use $\dot{\Gamma} = -\frac{1}{2}\dot{\gamma}$ in the equation (2.90)

$$T \equiv p_{22} - p_{11} = 6\eta\dot{\Gamma} . \qquad (2.104)$$

In strip biaxial stretching, using (2.72) ,

$$T \equiv p_{11} - p_{22} = 4\eta\dot{\varepsilon} \qquad (2.105a)$$

and

$$T^X \equiv p_{33} - p_{22} = 2\eta\dot{\varepsilon} . \qquad (2.105b)$$

If the stress T^X is not applied we get uniaxial stretching as has been verified experimentally [A11]; see the discussion in section 3.1.2 below. We can carry out the same sort of calculations as for uniaxial stretching and for the Newtonian fluid there is very little difference between the forms of stretching. If a constant force is applied to the edges of a square, we have (2.2a)

$$F = 6\eta\dot{\Gamma}LH$$

and using (2.30) and (2.31) and integrating gives

$$\frac{1}{L} = \frac{1}{L_o} - \frac{Ft}{6\eta V_M} \qquad (2.106a)$$

and $H = (\sqrt{H_o} - Ft/6\eta_o\sqrt{V_M})^2$ (2.106b)

with $L \to \infty$ and $H \to 0$ as $t \to 6\eta_o H_o L_o/F$ which is essentially the same as for uniaxial stretching. (The decrease in length is quadratic in time, t, as compared with a linear increase in length with t for uniaxial stretching.)

2.3.6 The co-rotational Maxwell model

For this model we have, from (2.89) with (2.79) and (2.80) the spinning equation

$$FU/Q + \lambda U d(FU/Q)/dX = 3\eta dU/dX . \quad (2.107a)$$

If we write $U_c = 3\eta Q/\lambda F$, a constant 'critical velocity', we can rearrange this to

$$(1 - U/U_c)\, dU/dX = U/\lambda U_c \quad (2.107b)$$

and see that this becomes singular ($dU/dX \to \infty$) if $U \to U_c$. We also see that if $U < U_c$ then $dU/dX > 0$ while if $U > U_c$ then $dU/dX < 0$ so that $U \to U_c$ whatever the values of U_o and the other parameters. We can obtain an expression for the breakdown distance for solutions; we find $U \to U_c$ as $X \to X_c$ where

$$X_c = \lambda\{U_c \log(U_c/U_o) + U_o - U_c\} . \quad (2.108a)$$

The corresponding time of existence is

$$\tau_c = \lambda\{U_c/U_o - 1 - \log(U_c.U_o)\} . \quad (2.108b)$$

For as long as solutions do exist we can calculate the ratio of the local stress to the local rate of strain and obtain

$$\frac{T(X)}{U'(X)} = \frac{FU(X)/Q}{U'(X)} = 3\eta\{1 - U(X)/U_c\} . \quad (2.109a)$$

This may be expressed, using (2.107b) to eliminate $U(X)/U_c$, in the form

$$T(X)/U'(X) = 3\eta/(1 + \lambda U'(X)) \quad (2.109b)$$

which gives a quantity with the dimensions of viscosity which is a function of material parameters and the local strain rate. Note however that this is <u>not</u> the elongational viscosity of this material, which is 3η - recall that elongational viscosity is defined for a steady uniform flow. The difference arises because, although the flow is spatially steady it is not steady for a particle of the liquid. The changing stress experienced by the particle leads to a different relation between instantaneous values of stress and strain rate because the constitutive equation involves the material time derivative of the stress.

In uniaxial stretching use of (2.89) gives

$$T + \lambda dT/dt = 3\eta\dot{\gamma} \ . \tag{2.110}$$

If a constant force is applied then (2.79) and (2.65) give the equivalent of (2.107b), namely

$$[1 - (\lambda F/3\eta A)] \, dA/dt = -F/3\eta \tag{2.111}$$

which has the same property of existence of solutions for a limited time. At constant rate of strain

$$T = 3\eta\dot{\gamma}(1 - e^{-t/\lambda}) \tag{2.112}$$

while if a constant stress is imposed

$$\dot{\gamma} = T/3\eta$$

and the viscoelastic nature of the model is not evident. However when the stress is applied at t = 0 there will be a jump in strain corresponding to the jump in stress, as illustrated in Figure 2.1. We discuss this, and use of integral models corresponding to (2.77) and (2.76), in section 5.2.1 below (see also sections 5.1.2, 5.2.4).

If the specimen is stretched with a constant velocity of its ends (relative to each other),

$$dL/dt = U \ , \tag{2.113a}$$

then the rate of strain is

$$\dot{\gamma} = U/(L_o + Ut) \ . \tag{2.113b}$$

In this case (2.110) has the solution

$$T(t) = (3\eta/\lambda)e^{-t/\lambda} \int_o^t U e^{t'/\lambda}/(L_o + Ut')dt' \tag{2.114}$$

which could be expressed in terms of the exponential integral function. The stress here has the predictable behaviour of rising from zero to a maximum value and then decreasing again as the rate of strain decreases.

In biaxial stretching the same sort of remarks apply to the co-rotational Maxwell model as to the Newtonian model, namely that because the elongational viscosity (stress and strain rate both independent of time) is constant (independent of strain rate or stress) the response to equal biaxial stretching is essentially the same as to uniaxial stretching. Instead of (2.104) we have

$$T + \lambda dT/dt = 6\eta\dot{\Gamma} \tag{2.115}$$

and similarly for strip biaxial stretching

$$T + \lambda dT/dt = 4\eta\dot{\varepsilon} \tag{2.116a}$$

and

$$T^X + \lambda dT^X/dt = 2\eta\dot{\varepsilon} \tag{2.116b}$$

2.3.7 The upper convected Maxwell model

The rate of strain (2.64) in (2.76) gives

$$p_{11} + \lambda(dp_{11}/dt - 2\dot{\gamma}p_{11}) = 2\eta\dot{\gamma}\ , \tag{2.117a}$$

$$p_{22} + \lambda(dp_{22}/dt + \dot{\gamma}p_{22}) = -\eta\dot{\gamma}\ . \tag{2.117b}$$

In order to eliminate both extra-stress components we subtract (2.117b) from (2.117a)

$$T + \lambda(dT/dt + \dot{\gamma}T) - 3\lambda\dot{\gamma}p_{11} = 3\eta\dot{\gamma} \tag{2.117c}$$

and eliminate p_{11} (which we cannot observe) by differentiation of (2.117c) and use of this with (2.117a) and (2.117c) to obtain the general equation

$$\lambda^2\ddot{T} + (2-\lambda\dot{\gamma}-\lambda\ddot{\gamma}/\dot{\gamma})\lambda\dot{T} + (1-\lambda\dot{\gamma}-2\lambda^2\dot{\gamma}^2-\lambda\ddot{\gamma}/\dot{\gamma})T = 3\eta\dot{\gamma} \tag{2.118}$$

where we are using the dot to denote differentiation with respect to time, t. It is not possible to solve this equation explicitly for most of the cases we wish to discuss here, and detailed consideration is postponed to chapter 5. In the case of constant rate of strain, when $\ddot{\gamma} = 0$, we can solve (2.118) and obtain

$$T = 3\eta\dot{\gamma}/(1+\lambda\dot{\gamma})(1-2\lambda\dot{\gamma}) + Be^{-(1+\lambda\dot{\gamma})t/\lambda} + Ce^{-(1-2\lambda\dot{\gamma})t/\lambda} \tag{2.119}$$

where the constants B and C are determined by the initial conditions. If the material is at rest for all time $t < 0$ we may assume that all stresses have relaxed and hence $p_{11}(0) = p_{22}(0) = 0$. This gives $B = -\eta\dot{\gamma}/(1+\lambda\dot{\gamma})$ and $C = -2\eta\dot{\gamma}/(1-2\lambda\dot{\gamma})$. The steady state solution gives

$$\eta_T = 3\eta/(1+\lambda\dot{\gamma})(1-2\lambda\dot{\gamma}) \qquad (2.120)$$

dependent on the rate of strain $\dot{\gamma}$ and in fact increasing with $\dot{\gamma}$, but the exponential terms only decay (so that a steady state is attained) if $-1 < \lambda\dot{\gamma} < \frac{1}{2}$. For the moment we shall consider positive values of $\lambda\dot{\gamma}$ only, since negative values correspond to equal biaxial stretching which we discuss below.

The constant stress equation is nonlinear in $\dot{\gamma}$, and we discuss its properties in section 5.2.3 below. As with any Maxwell model the upper convected model has an instantaneous jump in strain, but in addition the rate of strain is initially greater than the steady state value. The shape of the creep and recovery curves for the two models are shown in Figure 2.7, together with the Newtonian fluid for comparison. The graphs are graphs of

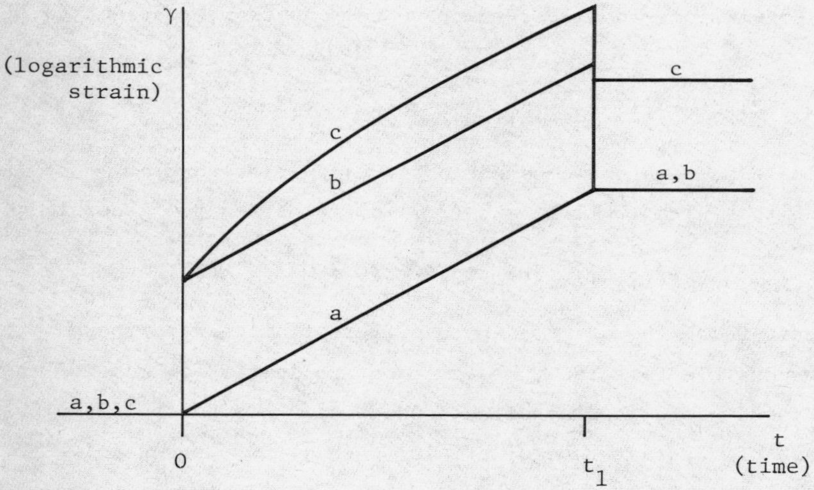

Figure 2.7 Creep and recovery curves: a - Newtonian, b - co-rotational Maxwell, c - upper convected Maxwell model.

strain γ defined by

$$\gamma(t) \;=\; \int_{0-}^{t} \dot{\gamma}(t')dt' \;=\; \log[L(t)/L_o] \qquad (2.121)$$

as a function of time, with a constant stress T applied at time $t_o = 0$ and removed at time t_1. The main features of the response are that for the co-rotational model the initial jump in strain (the elastic response) γ_e is equal to the strain recovered on removal of the stress γ_r and is linear in T, while for the upper convected model γ_e and γ_r are nonlinear functions of T, $\gamma_r < \gamma_e$ and γ_r is a decreasing function of t_1 taking values between γ_e for $t_1 = 0$ and a strictly positive limiting value for $t_1 \to \infty$. We shall see that this is not in agreement with experimental evidence.

The elongational viscosity of the upper convected model may alternatively be expressed as a function of the tensile stress

$$\eta_T \;=\; \frac{\eta T}{2G}\left\{1 + \frac{3G}{T} + \left[8 + (1 + \frac{3G}{T})^2\right]^{\frac{1}{2}}\right\} \qquad (2.122)$$

which is a monotonically increasing function of T, defined for all T, and it will further be shown (section 5.2.3) that steady flow is attained (at large times) for any value of the applied stress. If we apply a constant force we discover that solutions of (2.18) for $\dot{\gamma}$ tend to the limiting value $1/\lambda$ at large time, so that we have a flow with a constant rate of strain but an exponentially increasing stress. It is not obviously helpful to call this a steady flow, and the example highlights the difficulty of defining steady flows in useful ways as Lodge [L17, pp. 137,122] has pointed out.

The spinning equations we obtain, analogous to (2.117), are

$$p_{11} + \lambda(Udp_{11}/dX - 2p_{11}dU/dX) \;=\; 2\eta dU/dX, \qquad (2.123a)$$

$$p_{22} + \lambda(Udp_{22}/dX + p_{22}dU/dX) \;=\; -\eta dU/dX, \qquad (2.123b)$$

$$p_{11} - p_{22} \;=\; FU/Q\ . \qquad (2.123c)$$

We leave detailed discussion of this to chapter 6, and present here numerical results [D12] in terms of dimensionless parameters $\alpha = \lambda U_o/L$ and $\varepsilon = \eta Q/FL$. These may be interpreted as a dimensionless relaxation time and a reciprocal dimensionless force. In Figure 2.8 are graphs of $u(=U/U_o)$ against $x(=X/L)$ for various values of α and with ε chosen to give $u(1) = 20$. Computation

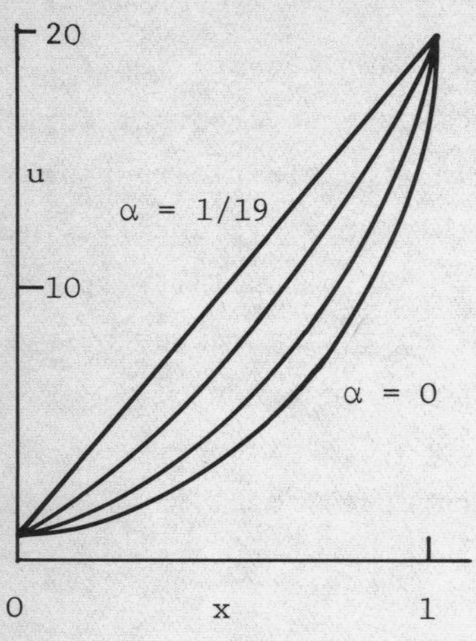

Figure 2.8 *Spinning velocity profiles.* D_R *fixed,* α *and* ε *varied.*

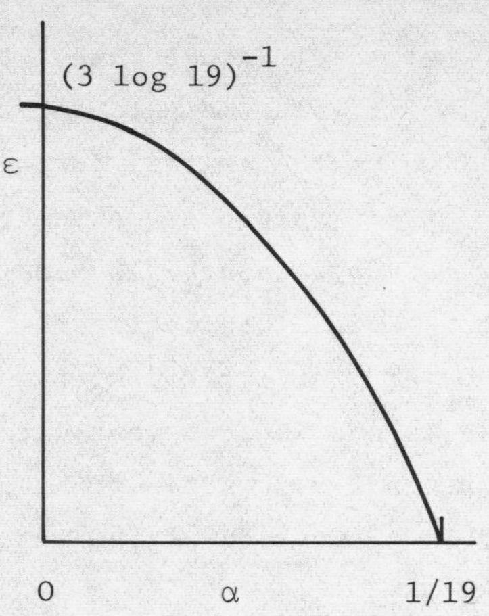

Figure 2.9 *Spinning force - relaxation time relation; force* ∝ *1/*ε*, relaxation time* ∝ α*,* D_R *fixed.*

and asymptotic solutions of the equations show that the initial condition on p_{11} needed to solve (2.123) does not have a very large effect on the solution. The results shown are obtained for $p_{11}(0) = FU_o/Q$ and hence $p_{22}(0) = 0$ by (2.123c). The point of interest here is that as α increases (i.e. visco-elasticity becomes more important) the value of ε has to decrease as shown in Figure 2.9, and reaches zero for the linear velocity profile shown for α = 1/19. This corresponds to an infinite force, and the conclusion is that however large a force we apply, we cannot get a draw ratio $D_R(= U(L)/U_o = u(1))$ larger than $1 + \alpha^{-1}$. The linear profile corresponds to a strain rate $u'(x) = \alpha^{-1}$, or $U'(X) = \lambda^{-1}$, which we have already found to be the maximum strain rate for constant force stretching. The large force limit may be demonstrated analytically and a perturbation expansion developed. This will be discussed below, together with a small α (nearly Newtonian) approximation.

It is not possible to get explicit expressions for the local ratio of stress to strain rate, but numerical computation shows that the value depends on the initial value of p_{11} as well as on the local strain rate and fluid parameters

λ and η. In the large force limit we find

$$\frac{p_{11}-p_{22}}{U'(X)} = \frac{FU_o}{Q}\left(\lambda + \frac{X}{U_o}\right) \tag{2.124}$$

which depends on the initial stress FU_o/Q and on position X, even though the strain rate is constant along the jet. The behaviour is in fact more solid-like than liquid-like, with stress proportional to strain, and underlines the dangers of using so-called apparent viscosities in interpreting spinning experiments.

The equations corresponding to (2.117) for equal biaxial extension (2.67) are

$$p_{11} + \lambda(dp_{11}/dt + 4\dot{\Gamma}p_{11}) = -4\eta\dot{\Gamma} \tag{2.125a}$$

$$p_{33} + \lambda(dp_{33}/dt - 2\dot{\Gamma}p_{33}) = 2\eta\dot{\Gamma} \tag{2.125b}$$

and the steady state viscosity for equal biaxial extension is

$$\eta_{eb} = \frac{T}{\dot{\Gamma}} = \frac{p_{33}-p_{11}}{\dot{\Gamma}} = \frac{2\eta}{1-2\lambda\dot{\Gamma}} - \frac{-4\eta}{1+4\lambda\dot{\Gamma}} = \frac{6\eta}{(1-2\lambda\dot{\Gamma})(1+4\lambda\dot{\Gamma})} \tag{2.126}$$

For positive values of $\dot{\Gamma}$ this decreases to a minimum value at $\dot{\Gamma} = 1/8\lambda$ before increasing without bound as $\dot{\Gamma}$ increases towards $1/2\lambda$, while for positive $\dot{\gamma}$ η_T increases monotonically with $\dot{\gamma}$. We may go on to obtain the analogue of (2.118) to be solved for stress given the kinematics, or for $\dot{\Gamma}$, given the stresses or forces applied.

Similarly we may analyse strip biaxial extension, governed by

$$p_{11} + \lambda(dp_{11}/dt - 2\dot{\varepsilon}p_{11}) = 2\eta\dot{\varepsilon} \tag{2.127a}$$

$$p_{22} + \lambda(dp_{22}/dt + 2\dot{\varepsilon}p_{22}) = -2\eta\dot{\varepsilon} \tag{2.127b}$$

$$p_{33} + \lambda dp_{33}/dt = 0 \tag{2.127c}$$

with applied stresses $T = p_{11}-p_{22}$ and $T^X = p_{33}-p_{22}$, and the steady state solution is

$$T = 4\eta\dot{\varepsilon}/(1-4\lambda^2\dot{\varepsilon}^2) \tag{2.44a}$$

$$T^X = 2\eta\dot{\varepsilon}/(1+2\lambda\dot{\varepsilon}) \tag{2.44b}$$

which tends to the Newtonian solution as $\dot{\varepsilon}$ tends to zero. We note finally that $T^X/\dot{\varepsilon}$ decreases as $\dot{\varepsilon}$ increases, so that the relative difference between strip biaxial and uniaxial stretching will decrease as the strain rate increases (since the difference is associated with the additional constraining stress, T^X).

In conclusion we remind the reader that the predictions obtained with these very simple models will rarely be quantitatively correct. The aim is to develop a qualitative understanding of the way different features of constitutive equations influence the predictions obtained using them. More quantitative success can be obtained in many ways - by replacing constant parameters by functions of the stress or strain rate, by adding different terms as in going to the Jeffreys and Oldroyd models, or by introducing a spectrum of relaxation times. We discuss these various approaches below (in chapters 5 and 6).

3 Experimental methods and results

"Another school declares ... that the universe is comparable to those cryptographs in which not all the symbols are valid."

Jorge Luis Borges.

At the time of writing two reviews are about to be published which offer the reader alternative views of this topic. Both Dealy [D 7] and White [W17] deal exclusively with polymer melts, while here we also refer to some results for elastomers and, in section 3.3, polymer solutions. Other materials such as molten glass have been mentioned in the historical survey above (section 1.1). However the major research effort in recent years has been devoted to polymer melts and these are our main subject here. We shall consider here only our three basic flows and some closely related flows. An obvious omission is that of converging flow, which we discuss in section 7.1 below. We take the view that at present its interpretation is less well-founded in theory, although it provides a simple technique of discovering something about the elongational response of a liquid.

3.1 UNIAXIAL STRETCHING

3.1.1 Experimental methods for polymers

A wide variety of techniques have been employed, and considerable ingenuity exercised, in arranging "tensile tests" for molten and almost molten polymers, and even the best of those currently used have limitations, notably on the viscosity of material with which they may be used. For high viscosity materials (viscosities of 10^5 to 10^9 Pas) there is by now a reasonable amount of data published, notably on molten polyethylene, molten polystyrene and polyisobutylene, but for lower viscosities other methods appear to be necessary. The study by Dealy and co-workers [D 6] illustrates the difficulties in designing a constant-stress "melt extensiometer" to deal with materials of lower viscosities (10^4 Pas), difficulties associated largely with apparatus friction and with fluid dynamic drag. A supporting fluid (in which to float the test specimen) is necessary to minimize deformation of the material under its own weight.

The popular technique for arranging a test at constant stress is to apply a constant force to a suitably shaped cam, so that the force applied to the specimen is inversely proportional to its length and hence directly proportional to its area. This was first done for polymer melts by Cogswell [C24,C25] though Leaderman [L 6] developed such an apparatus for work on rubbery solids. Several other workers have developed Cogswell's design or independently arrived at essentially the same solution to the problem [V 4,N 2,B 2,M37,L 3] and details of cam design are given by Dealy [D 6]. The alternative of using a servomotor regulated by signals from force and displacement measuring devices is described by Meissner [M21] (essentially as a development of his constant strain rate apparatus) and proposed by Dealy [D 6]. However in spite of Meissner's success, no further use of his apparatus for constant stress testing has been reported, and Dealy has not yet reported successful use of his latest instrument [R6] in this way. The apparatus developed by Münstedt appears to be the most highly developed [L 3] and is the best constant stress apparatus constructed and working at the time of writing. It overcomes [D 7] most of the limitations mentioned above [D 6]. It is a strong indication of the industrial importance of this work that major developments have taken place in the laboratories of BASF in Germany (Meissner, Münstedt, Laun) and ICI in England (Cogswell).

The other rheologically satisfactory test is at constant strain rate, and there is more variety here in solutions to the problem of controlling the velocity with which the ends of the specimen are pulled apart so that $L^{-1}dL/dt$ is constant while the length of the specimen (L) increases (to perhaps forty times its original length (L_o)). The first published account here is again from an industrial laboratory, by Ballman [B 3] of Monsanto in the U.S.A.. He used a tensile testing instrument with the velocity controlled by a pre-programmed function generator. Vinogradov [R 1 ,V 2 ,V 3], Stevenson [S 18] and Peng and Landel [P 7] adopt essentially the same technique with a function determined to give the desired constant strain rate for the monitored section of the specimen. The use of a cam whose shape determines the position of the end of the specimen as a function of time by Metzner and co-workers [A 11] is a further example of the same (open-loop) approach to controlling the strain rate, and Dealy [D 7] points out the advantage of this technique, that a given length L(t) determines a strain rate at t, while a given velocity U(t) does not since the strain rate is U(t)/L(t) and L(t) is only determined correctly if U(t) has followed the correct function since t = 0 (implying a jump in U at t = 0). The history of this development does highlight the need for careful and thoughtful experiments and it must still be recognised that a high degree of skill is required to obtain good data on elongational flow. Evidence of this may be found by comparing an earlier account [R 7] with [A11]; in the oral presentation of the former Metzner reported the later results and emphasised the importance of temperature control. (For polystyrene a change of 3°C causes the viscosity to change by a factor of two.) We shall return to this work later, when we discuss the results of experiments, remarking now only that the full published account [A11] does appear to contain valuable and reliable experimental results.

This apparatus has also been used with an arbitrary non-constant strain rate, which we discuss below. Shaw [S 6] has similarly carried out both constant strain rate tests and tests with an arbitrary chosen strain rate history, again using open-loop control. The novel feature of his work is the use of ring specimens rather than cylindrical specimens. This avoids problems of clamping specimen ends, since the method of extension is just that of stretching an elastic band, and in fact this type of specimen has been used in testing elastomers. De Vries and Bonnebat [V12], as well as Dealy [R6], have recently developed devices employing closed-loop (feedback) control to maintain a constant strain rate. The former have the sample in a gas rather than supported in a liquid and so are restricted to more viscous materials, but the basic ideas seem similar.

A fundamentally different approach is adopted by Meissner [M19,M20,M21] who extends a specimen between two pairs of gears or "rotational clamps". These are a fixed distance apart, so the length of specimen under test remains constant, and each end of the specimen is drawn through the gears at a constant velocity, so that the elongation of the material between the pairs of gears takes place at a constant strain rate. This technique avoids the problem of needing a very long apparatus to get large strains or (equivalently) experiments of long duration, and in the form developed by Laun [L 3,L 4] the apparatus can give a logarithmic strain of 6, corresponding to an extension ratio of 400. The limitations are that the specimen should remain reasonably uniform, and that the force which must be measured should not get too small. There is a careful discussion of these factors by Laun and Münstedt [L3], and here we have again some reliable results which are the more valuable because of the capability of extending to the large strain of 6.

There are developments of this idea which have the attraction of simplicity, but which are not apparently capable of giving results which are as accurate as the Meissner apparatus. The results of Macosko and Lorntson [M 8] or of Everage and Ballman [E 9] are not to be ignored however - there is little enough data on elongational flow and the trade-off between high accuracy and the possibility of carrying out elongational tests as part of a routine evaluation of a polymer is a very fair one. The apparatus used is a sophisticated rotational viscometer (the Rheometrics Mechanical Spectrometer) whose rotating spindle is used to wind up one end of the specimen and is also raised so that the specimen does not wrap round itself but is taken up on the spindle and hence at constant velocity. The other end of the specimen is held fixed so that we have essentially half of the Meissner arrangement. Garfield, Pearson and Connelly [G 1] have recently reported that in this apparatus the specimen extension is not uniform and according to Dealy [D 7] they have shown that it is not suitable for large strain extensions. In fact Pearson [P 2] finds that for polystyrene the nonuniformity of the stretching is most marked at low strain rates (e.g. .006 s^{-1}) and at high temperatures (e.g. 155°C), where the behaviour should be approaching Newtonian behaviour. Under these conditions calculation of strain and strain rate from the overall length of the specimen is not reliable for strains of 0.5 or more. However at higher strain rate (0.04 s^{-1}) and lower temperature (132°C) a strain of about 3 was attained with reasonably uniform stretching of the specimen. Dealy also reports that the manufacturers of this apparatus (Rheometrics Inc.) are planning to manufacture an apparatus based more closely on Meissner's design. The nonuniformity is observed to be more pronounced with some polymers (e.g. polystyrene) than others (e.g. low density polyethylene). Most careful workers at the very least verify that the specimen diameter at the end of the experiment is sufficiently near to uniformity [A11,P 2 ,L 5].

The most recent apparatus developed by Ide and White [I3 ,I4] is very similar to that of Macosko and Lorntson, but uses a driven take-up roll or drum pulling one end of the specimen and the other end is held fixed. In an earlier study of spinnability [I2] where force measurements were not made two drums rotating in opposite directions were used, but the convenience of measuring the tensile force on a fixed end clearly led to the development reported (see also [L 5]). Ide and White report on a large number of experiments on different polymers [I 4] with the main limitation, as with almost all workers, of a total logarithmic strain of about three. It is also important to note that their measurements are taken right up to specimen breakage, so that nonuniformity of specimens at the highest elongation reported seems likely. (Since the breakage of extending filaments is extremely important, and is the main subject of study in two of the three papers by Ide and White, this is not a criticism of the work they report, but merely a note of caution in regard to its limitations.)

Finally we should mention the experimentally simpler techniques of applying a constant force or a constant velocity. In conventional testing of solids the distinctions are not important because for an elastic material the mode of testing is irrelevant, and because the strains are so small that the error involved in replacing stress by "engineering stress" or force divided by initial area is very small, as is the error in taking the rate of strain to be velocity divided by initial length of the specimen. Great care must be taken in reading accounts of work where this background influences authors and the distinctions are not clearly made for deformations which are not small.

It is also important to carry out measurements on recovery after removing applied stresses or forces, and there is a limited amount of data here which we review below. The basic technique is to cut the specimen and measure the retraction of parts of it at the temperature of the test. Detailed discussion of this measurement, and of the possible influence of surface tension on the elastic recovery is given by Vinogradov [V 1,V 2]. Meissner, Laun and Münstedt do not go into detail on their measurements of recovery, and it seems, from the agreement found between two techniques by Vinogradov, that the details of the technique do not need to be specified in precise detail. The influence of surface tension during the stretching is discussed below (section 5.3.1).

3.1.2 Experimental results for polymers

Ide and White have summarised quantitative experimental studies of isothermal elongational flow of polymer melts [I 4, table I] including both stretching and spinning experiments. Here (Table 3.1) we attempt to present a slightly more detailed picture for stretching experiments. It should be emphasised that the classification of the polymers is somewhat crude, in the sense that there may be considerable differences between different polystyrenes or polyethylenes, and polyisobutylenes of different molecular mass are described as liquids and as elastomers, and may even be crosslinked. The coverage of elastomers in Table 3.1 (of SBR and natural rubber in particular) is not comprehensive, but merely illustrates the sort of data that is available. The aim of the table is to assist the mathematician seeking data for comparison with theoretical predictions. It also serves to highlight (by omission) polymers and measurements where experimental work might be of value and (by their inclusion) the polymers for which experiments are easier or data are in demand. It seems fair to compare the large number of entries for polyethylene, polystyrene and polyisobutylene with the major polymers (in terms of tonnage produced annually) where the 'big three' are the polyolefines (polyethylenes and polypropylene), polyvinyl chloride, and polystyrene (with its copolymers such as SBR). The popularity of polyisobutylene is certainly attributable to ease of experimentation, but otherwise the data reflect production quite well. Notable absences are the major polymers used in artificial fibre production (nylons, polyesters, polyacrylonitrile) which we shall not even see well represented in our later discussion of published accounts of spinning experiments. (The two entries for polyethylene terephthalate are in connection with studies of orientation and drawing of polymer fibres and sheets.) Part, at least, of the explanation for this is that these polymers are commonly spun at higher temperatures than are possible in present experimental arrangements, and also are less viscous so that force measurement is more difficult.

In this section we shall not attempt to discuss all the available data (see the review by Cogswell [C27] for elongational viscosity data), but shall select some in order to present the major features of the response of polymer melts to uniaxial stretching flows. The flows in Table 3.1 were obtained by stretching both cylindrical and flat specimens, and it is important first to note that the kinematics are the same whether there is axial symmetry or not. This has been experimentally verified, notably by Metzner and co-workers [A11] who stretched specimens with rectangular cross-sections of initial dimensions 25mm × 2.5mm and verified that the longitudinal strain rate equalled minus one half of the lateral strain rate. This also confirms directly the assumption of incompressibility for these flows (for their materials at least), and the method of measurement also provided evidence of uniformity of the deformation.

We obtain two types of information from measurements of stress, strain and strain rate. If steady flow is attained, with stress and strain rate reaching constant values then their ratio gives the elongational viscosity, which is a material property. This may be measured and recorded as a function of elongational stress or strain rate. The second type of information is obtained from unsteady flows where the stress or the strain rate (or both) are functions of time. In this case we do not obtain a material property directly, since the ratio of the time-dependent stress to strain rate depends on the mode of stretching as well as on the material and, as well as on the time since the start of the experiment, it may depend on the deformation history of the material before the experiment. It is possible to obtain a function of time which is a material property if we define the experiment completely. Meissner [M20,M21] follows Giesekus [G10] in using the term stressing viscosity (Spannviskosität) for the ratio of stress to strain rate in an experiment at constant strain rate on a material which has been at rest for a long time before the start of the experiment (so that it is stress-free or fully relaxed and has no memory of any previous deformation). This material function may depend on the strain rate for a nonlinear material, as well as depending on the time or on the strain (which is the time multiplied by the constant strain rate).

It is extremely important not to confuse elongational viscosity with stressing viscosity, or with the ratio of stress to strain rate for any different experiment (e.g. the reciprocal of the creep compliance in a creep experiment). Note that it involves a contradiction to talk of elongational viscosity as a function of strain since, if this were the case, at constant strain rate the strain grows with time so that the elongational viscosity would change, and hence the stress would change and a steady state would be impossible. This is not to say that some empirical correlations may not be found between a ratio of stress to strain rate and strain, or indeed that such correlations may not be useful if treated with caution, but the need for careful thought must be emphasised.

There is a further possibility for serious confusion in dealing with elongational viscosity, which arises from a different definition of this quantity. Following Kargin and Sogolova [K 6] Vinogradov and co-workers

[R1,V1,V2,V3,V4,V5,V7] measure recovery after the removal of the stress and split the strain into a recoverable or elastic part and a non-recoverable, irreversible or viscous part (equal to the strain when the specimen has completely recovered). Since this can be done after any time the two parts of the strain may be measured as functions of time, and the viscous strain rate calculated (as the rate of increase of the non-recoverable strain). Then the quantity stress divided by viscous strain rate is calculated, and is called the elongational viscosity. We shall not use this term here for this quantity since, whatever the merits of this approach, the use of one name for two different quantities is bound to lead to confusion (and has indeed done so). It has been inferred, quite wrongly, from graphs in some of these papers [V2,V3] that the stress in uniaxial elongation at constant strain rate goes through a maximum before settling down to a steady value (in a manner analogous to the well-established "stress overshoot" in the start-up of simple shear). The quantity which does go through a maximum is the ratio of stress to viscous strain rate, both of which are time-varying quantities. Recalculation of the data from [V3] by Vinogradov [V10] shows that in fact the stress increases monotonically with time in the experiments at constant strain rate. An alternative confusion with this collection of data is that constant strain rate and constant velocity data are presented on the same diagrams [R1,V2], and are referred to as constant deformation rate and constant extension rate respectively. As we have seen (section 2.3.6) constant velocity stretching for a simple viscoelastic model must involve a maximum in the stress as a function of time. This conclusion is probably true for any viscoelastic material and is also borne out by all available experimental evidence.

Having distinguished between the two types of information we may obtain, the next question is whether we can always obtain both. Stressing viscosity poses no problem in principle, but steady elongational flow may not be possible, so that an elongational viscosity may not exist and if the steady state is possible in principle it may not be reached in any given experiment. This corresponds to the theoretical situation for the upper convected Maxwell model, which can be nicely interpreted by saying that at low strain rates the model response is liquid-like with steady flow (constant strain rate) being sustained by a constant stress while at high strain rates the model response is solid-like, with stress increasing with strain and never reaching a steady value since the strain increases with time at a constant rate. There has been some discussion and disagreement concerning this type of behaviour, since the interpretation in terms of an infinite elongational viscosity, and of a critical strain rate above which steady stretching was impossible is a little hard to accept. As we have seen this interpretation is an over-simplification and the transient behaviour needs to be considered, giving our interpretation of the critical strain rate as the rate of stretching below which response is liquid-like and above which it is solid-like.

It is extremely difficult to produce experimental evidence as to whether this aspect of the response of the upper convected Maxwell model is a reasonable representation of real material behaviour. The absence of a steady state may merely mean that the experiment has not been carried on for long enough to reach the steady state. On the other hand the existence of steady states may be taken to show that the critical strain rate has not been exceeded. Hence the tabulation of what evidence there is [P16] is not a very useful exercise, since such questions are not decided democratically by a majority vote, but by the best available evidence. The results of Laun [L3, L4] provide a good example of the attainment of a steady state which did not appear to exist, since they extend the earlier work of Meissner to longer times (and greater strains). Meissner's results [M20] have been taken as good evidence of the qualitative correctness of the rubberlike liquid model [C8] which has the same response as the upper convected Maxwell model. With logarithmic strains of up to four, it was found that at all but the lowest strain rate (0.001 s^{-1}) the stress grew in a manner suggesting that it would continue to grow without bound. Laun has now shown that for some polymers including one that Meissner used [M22] if the stretching is continued to a strain of six (an extension ratio of four hundred) the stress reaches a steady value. In fact the stress reaches its steady value either if the time of elongation exceeds the greatest relaxation time of the material or if the strain exceeds four (so that from strains of four to six the stress is seen to be constant).

Evidence for the rejection of the contrary hypothesis, that strain rates are not great enough to get truly solid-like behaviour, while not completely convincing, is obtainable from independent estimates of relaxation times of the material, and the best guess at present is that unbounded stress growth is not found in the elongation of polymer melts at constant strain rate.

This is not to say that at short times (as in the Meissner experiments) the Maxwell model may not give a useful description of material response, and of strain hardening behaviour in particular.

When we come to compare behaviour in elongation with behaviour in shear we find that the elongational viscosity at low strain rates approaches three times the viscosity (as measured in shear flows at low strain rates). At higher strain rates, in contrast to behaviour in shear where almost all polymers melts are shear thinning (viscosity decreases with increasing shear rate), the behaviour varies between different polymers. For molten low density polyethylene all the evidence shows that the elongational viscosity increases with increasing strain rate initially. The best evidence [L4] is that at high strain rates elongational viscosity decreases with increasing strain rate, but it should be said that there is a limited amount of reliable data for high strain rates. The other common polymer melts do not show the maximum in elongational viscosity, and either show little influence of strain rate or have elongational viscosity decreasing with increasing strain rate. The latter behaviour, typical of linear polymers such as high density polyethylene and polypropylene, appears to be associated with greater difficulty in carrying out stretching experiments or poorer spinnability (though this is not the only factor relevant to spinnability). We should note here that polymer solutions show qualitatively different behaviour, with much larger ratios of elongational viscosity to shear viscosity, and strongly increasing elongational viscosity as a function of shear rate. The evidence for this is less direct since we cannot carry out straightforward tensile tests on these much less viscous liquids, and we discuss the experiments below (spinning, 3.3; converging flows, 7.1). One useful idea due to Wagner [W 2] is that, rather than thinking of the elongational response of low density polyethylene as strain rate hardening (elongational viscosity increasing), we can compare its response with that of the upper convected Maxwell model. Relative to this model the material response is strain rate thinning both in elongation and in shear (where the model has a constant viscosity). This is given quantitative significance in the work of Metzner and colleagues [A11] whose date for polystyrene is described by an upper convected Maxwell model with different parameters chosen for each constant strain rate. The elongational viscosity η_T (2.46) is given in terms of functions η and λ (viscosity coefficient and relaxation time) which depend on the strain rate $\dot{\gamma}$, and it is found that these both decrease with increasing strain rate.

Most important is the fact that this dependence, when expressed in terms of the second invariant of the rate of strain tensor is the same (quantitatively) as the dependence found for the same parameters from shear flow measurements. Thus, by comparison with the upper convected Maxwell model (with constant coefficients) the polystyrene melt shows the same degree of strain rate thinning in shear and in elongation. This is a quantitative demonstration of the validity of Wagner's idea and shows that, at least for small strain rates (and for small strain at large strain rates) the variable-parameter Maxwell model has definite value as a realistic simple model of material response. Note that this determination of Maxwell parameters was done with the transient data (from stressing experiments - constant strain rate), and steady flow conditions were not achieved for the higher strain rates used. It is interesting to see what the model predicts for these higher strain rates if larger strains could be achieved. The condition $\lambda \dot{\gamma} < \frac{1}{2}$ for the existence of steady flow (from 2.19) holds if λ is a function of $\dot{\gamma}$ as well as if it is a constant. Now the data for larger values of $\dot{\gamma}$ [A11, Figure 11] in fact has $\lambda\dot{\gamma}$ very close to and generally larger than $\frac{1}{2}$, so we cannot really draw any strong conclusion from the results. Certainly it will not require a large change in material behaviour at large strains to get steady flow, as found in polyethylene [L 3,L 4].

The results of these workers [A11] for low density polyethylene deserve some attention, as being different from polystyrene and not, in fact, bearing out Wagner's idea quantitatively. They find the Maxwell response for strains of up to about 1.3, and then observe strain hardening. This sort of behaviour might be expected from a model with several relaxation times, an idea that we discuss below. We note here that Chang and Lodge [C8] had more success in fitting Meissner's data using five relaxation times, obtaining the strain hardening response, but only moderately good quantitative agreement was obtained, with the model overestimating the stressing viscosity.

Finally we consider measurements of recoverable strain which give us important additional data for testing model predictions. The measurements of Laun and Münstedt [L 3] show that creep and stressing experiments are consistent with each other, and the basic behaviour is a smooth increase of recoverable strain from zero at the start of an experiment to a constant value whose magnitude depends on the stress, and which is reached when the flow becomes steady. In steady flow our definition of elongational viscosity and that of Vinogradov are the same, and it is interesting to note, perhaps as a point in favour of the latter, that it appears that a graph of stress divided by viscous strain rate against strain may be extrapolated back to a value of three times the zero shear viscosity at zero strain. This experimental observation has not been explained in detail in terms of any theory - the factor of three is not strictly speaking the Trouton ratio, since this refers to steady flow, but it appears to be something analogous, suggesting an initially linear response with nonlinear (strain hardening) behaviour at later times. One obvious feature of the upper convected Maxwell model is that the recoverable strain is initially a maximum and decreases to a steady value [P16]. We shall discuss this and other theoretical predictions below.

Date	*Investigator*	*Mode of stretching (constant -)*	*Recovery measured?*	*Maximum strain (logarithmic)*
POLYSTYRENE				
1938	Jenckel & Uberreiter [J5]	Force	Yes	1.1
1964	Karam & Bellinger [K4]	Force	Yes	-
1965	Ballman [B3]	Strain rate	No	0.25
1969	Karam, Hyun & Bellinger [K5]	Force	Yes	0.7
1970	Vinogradov, Fikhman, Radushkevich & Malkin [V3]	Strain rate, strain†	Yes	2.3
1972	Vinogradov, Fikhman & Radushkevich [V4]	Stress	Yes	2.7
1975	Kamei & Onogi [K2]	Velocity	Yes	3.
1975	Lai & Holt [L1]	Velocity	No	1.4
1975	Münstedt [M37]	Stress	Yes	2.5

† stress relaxation measured

Table 3.1 Experimental investigations of uniaxial extension.

Date	Investigator	Mode of stretching (constant -)	Recovery measured?	Maximum strain (logarithmic)
1975	Takaki & Bogue [T 1]	Force	No	3.
1976	Everage & Ballman [E 9]	Strain rate	No	3.
1976	Ide & White [I 4]	Strain rate	No	5.
1976	de Vries, Bonnebat & Beautemps [V13]	Strain rate	No	1.
1977	Agrawal, Lee, Lorntson, Richardson, Wissbrun & Metzner [A11]	Strain rate	No	5.
1977	Garfield, Pearson & Connelly [G 1, P 2]	Strain rate	No	3.
1978	Chan, White & Oyanagi [C 5]	Strain rate	No	3.
LOW DENSITY POLYETHYLENE				
1968	Cogswell [C24, C25]	Stress	No	1.8
1969	Meissner [M19, M20, M23]	Strain rate	Yes	4.
1972	Meissner [M21]	Strain rate*, stress	Yes	4.
1972	Chen, Hagler, Abbott, Bogue & White [C14]	Force	No	2.6
1976	Ide & White [I 4]	Strain rate	No	11.
1976	Laun & Münstedt [L 3]	Strain rate, stress	Yes	6.
1976	Rhi-Sausi & Dealy [R 6]	Strain rate	No	2.5
1976	Shaw [S 6]	Strain rate*	No	2.9
1977	Agrawal, Lee, Lorntson, Richardson, Wissbrun & Metzner [A11]	Strain rate*	No	3.
1977	Laun [L 4]	Strain rate	Yes	6.
HIGH DENSITY POLYETHYLENE				
1969	Meissner [M19]	Strain rate	No	3.

* arbitrary strain rate histories also used

Table 3.1 (continued)

Date	Investigator	Mode of stretching (constant -)	Recovery measured?	Maximum strain (logarithmic)
1972	Chen, Hagler, Abbott, Bogue & White [C14]	Force	No	No measurable strain possible without filament breakage
1973	Macosko & Lorntson [M 8]	Strain rate	No	2.2
1976	Ide & White [I4]	Strain rate	No	3.5
1976	Shaw [S6]	Strain rate*	No	2.9
1977	Laun [L4]	Strain rate	Yes	6.
1978	Chan, White & Oyanagi [C 5]	Strain rate	No	2.
POLYISOBUTYLENE				
1949	Kargin & Sogolova [K8]	Force	Yes	2.
1952	Dahlquist & Hatfield [D1]	Stress	No	0.7
1956	Smith [S10]	Velocity	No	0.9
1965	Zapas & Craft [Z 3]	Force, velocity	Yes	0.7
1968	Radushkevich, Fikhman & Vinogradov [R 1]	Strain rate, velocity	Yes	2.3
1969	Vinogradov, Leonov & Prokunin [V1]	Velocity	Yes	1.3
1969	Goldberg, Bernstein & Lianis [G13]	Strain rate	No	1.
1969	Middleman [M31]	Strain†, velocity	No	4.
1970	Vinogradov, Radushkevich & Fikhman [V 2]	Strain rate, velocity	Yes	2.3
1972	Stevenson [S18]	Strain rate	No	3.
1973	Peng & Landel [P 7]	Strain rate	No	2.
1974	Baily [B 2]	Stress	No	-

* arbitrary strain rate histories also used
† stress relaxation measured

Table 3.1 (continued)

Date	*Investigator*	*Mode of stretching (constant -)*	*Recovery measured?*	*Maximum strain (logarithmic)*
1974	Cogswell & Moore [C26]	Stress	No	1.2
1976	Djiauw & Gent [D22]	Strain†	Yes	1.
1976	Taylor, Greco, Kramer & Ferry [T14]	Strain†	No	0.7
1977	Acierno, la Mantia, de Cindio & Nicodemo [A 7]	Velocity	Yes	1.
POLYETHYLENE TEREPHTHALATE				
1976	de Vries, Bonnebat & Beautemps [V13]	Strain rate	No	1.8
1976	Titomanlio, Anshus, Astarita & Metzner [T20]	Velocity	(Yes)	4.5
POLYPROPYLENE				
1969	Cogswell [C25]	Stress	No	Necking prevents viscosity measurement
1976	Ide & White [I 4]	Strain rate	No	4.
1976	de Vries, Bonnebat & Beautemps [V 13]	Strain rate	No	1.5 (necking reported)
POLYMETHYL METHACRYLATE				
1955	Bueche [B28]	Force	Yes	0.1
1969	Cogswell [C25]	Stress	No	-
1970	Williams [W22]	Velocity	No	0.9
1975	Lai & Holt [L 1]	Velocity	No	1.4
1976	Sadd & Morris [S 1]	Velocity, strain†	No	0.06
1976	Ide & White [I 4]	Strain rate	No	5.

† stress relaxation measured

Table 3.1 (continued)

Date	Investigator	Mode of stretching (constant -)	Recovery measured?	Maximum strain (logarithmic)
POLYVINYLCHLORIDE				
1962	Leaderman [L 6]	Force, stress	Yes	0.24
1966	Zapas & Craft [Z 3]	Force, velocity	Yes	0.7
1967	Findley & Lai [F 7]	Stress	Yes	0.01
1972	Vinogradov, Fikhman & Alekseyeva [V5]	Strain rate, velocity	Yes	2.7
1976	de Vries & Bonnebat [V12]	Strain rate	No	1.8
1976	Kazama, Sumida, Ogawa, Okai & Hiraoka [K14]	Strain rate	Yes	1.
POLYISOPRENE				
1972	Stevenson [S18]	Strain rate	No	2.4
1974	Djiauw & Gent [D22]	Strain†	Yes	1.
1975	Vinogradov [V 7]	Strain rate	Yes	2.5
EPOXY RESIN				
1976	Smith [S14]	Velocity†	No	0.1
ETHYLENE-PROPYLENE COPOLYMER				
1975	Greco, Taylor, Kramer & Ferry [G19]	Strain†	No	1.1
1975	Kramer, Taylor, Ferry & McDonel [K17]	Strain†	No	0.6
NATURAL RUBBER				
1951	Rivlin & Saunders [R 9]	Strain	-	1.3
1975	Jones & Treloar [J 8]	Strain	-	1.
SBR (STYRENE-BUTADIENE RUBBER)				
1952	Dahlquist & Hatfield [D 1]	Stress	No	1.1

† stress relaxation measured

Table 3.1 (continued)

Date	*Investigator*	*Mode of stretching (constant -)*	*Recovery measured?*	*Maximum strain (logarithmic)*
1958	Smith [S11]	Velocity	No	1.8
1968	Smith & Frederick [S12]	Velocity	No	1.4
1969	Smith & Dickie [S13]	Velocity	No	2.
1969	Goldberg, Bernstein & Lianis [G13]	Strain rate	No	1.
1970	McGuirt & Lianis [M3, M4]	Strain†, strain rate, velocity	No	1.
1970	Flowers & Lianis [F11]	Strain†, velocity	No	1.
1974	Djiauw & Gent [D22]	Strain†	Yes	1.
POLYBUTADIENE				
1975	Vinogradov [V7]	Strain rate	Yes	2.5
1975	Cohen & Carruthers [C29]	-	-	-
1976	Noordermeer & Ferry [N11]	Strain†	No	0.7
FILLED POLYMERS				
BEAD-FILLED STYRENE ACRYLONITRILE COPOLYMER				
1974	Nazem & Hill [N2]	Stress	No	0.6
GLASS GIBRE REINFORCED HIGH DENSITY POLYETHYLENE				
1978	Chan, White & Oyanagi [C5]	Strain rate	No	1.
GLASS FIBRE REINFORCED POLYSTYRENE				
1978	Chan, White & Oyanagi [C5]	Strain rate	No	1.

† stress relaxation measured

Table 3.1 (continued)

3.2 BIAXIAL STRETCHING

3.2.1 Experimental methods for polymers

The techniques used are drawn largely from work on solid polymers and elastomers, and are applicable only to very viscous materials. Bubble inflation is the most popular technique, following Treloar [T22], giving a good approximation to equal biaxial extension near the pole of a spherical cap bubble. This has been used for polyisobutylene by a number of authors and by me or two authors for other polymers. Strip biaxial or two-dimensional extension is more difficult to realise and three techniques may be noted. The first of these is the extension of a rectangular sheet held along its four edges as described above, which was used (along with other methods) with notable success by Rivlin and Saunders [R9] on natural rubber, but has not proved popular for thermoplastics. The extension of a cylindrical tube with an internal pressure applied to keep the cylinder radius constant has also been used successfully, by Smith and Frederick for a styrene-butadiene rubber (SBR) [S12], and is an elegant solution of the problem of providing a constraint in the transverse direction. Denson and Crady [D18] have adapted the bubble inflation technique by inflating a rectangular sheet (with sides one inch and four inches), and found that the extension in the direction parallel to the long side is less than 1% of the extension in the direction parallel to the short side so that the sheet inflates to a cylinder. This latter technique appears to be the only one used to date on thermoplastic materials.

The bubble inflation technique for equal biaxial extension of thermoplastics has also been developed by Denson and colleagues from a gas driven inflation apparatus designed to give an approximately constant stress (by exploiting the kinematics and the compressibility of the gas) [D16,J12,J13] to a liquid driven apparatus with closed-loop feedback control able to give constant stress, constant strain rate or in fact any desired stress or strain history [D19]. A similar development has been discussed by Dealy [D5,D8]. The apparatus of Baily [B2] is similar to the early [D16] Denson apparatus and in particular Baily does not measure local deformation near the pole of the bubble as is done in later work [J12,J13,M9,V12,V13] but was an average value. This does introduce some errors as the careful discussion of Joye, Poehlein and Denson [J12] shows, and discrepancies between results [D16,J12] can be explained entirely on this basis [J13]. It appears that the assumption of a uniform spherical bubble leads to an underestimate of the strain rate and of the stress, and the overall effect of this will depend on the material.

The experiments of Maerker and Schowalter [M9] attempt to obtain a constant strain rate by regulating gas flow to the inflating bubble. In addition to measuring local deformation, this work is also noteworthy because the polyisobutylenes used have lower molecular weights and viscosities than any of the other materials used in biaxial experiments. With zero shear rate viscosities of 10^5 Pas, it seems that the lower limit for this type of experiment has been reached, although Dealy [D5] suggests that the use of a liquid for inflation will also give the necessary support to less viscous materials.

The use of closed-loop feedback has also been developed by De Vries and Bonnebat [V12,V13], who use a gas-driven bubble inflation apparatus. The controlled quantity is strain, made to increase linearly with time by regulating the gas pressure, and the strain is measured directly by a strain gauge attached to the polymer sheet across the pole of the bubble. This ensures that it is the local strain rate which is kept constant. We should note here that the previously mentioned closed loop systems [D19,D5] use average values of quantities, based on measurement of pressure and volume of the bubble. This will lead to inaccuracies in equal biaxial measurements, which should not be ignored. However Denson and Crady [D18] observe in strip biaxial extension, where a cylindrical bubble is being inflated, that the deformation is uniform across the bubble. Indeed they point out that the kinematic constraint which prevents uniform equal biaxial extension is irrelevant in the cylindrical geometry and (apart from end effects) there is no reason why the deformation should not be uniform. With a spherical bubble, extensions will be the same in all directions in the surface at the pole, but at the edge of the bubble extension is only possible in the azimuthal direction and not in the transverse direction (around the fixed edge) so equal biaxial extension is only possible well away from the fixed edge. In their recent work Denson and Hylton [D19] refer to both geometries but present results only for strip biaxial extension. Dealy [D8] observes that (as noted earlier by Joye [J13]) in the inflation of a spherical bubble the strain rate at the pole is proportional to the average strain rate, so that closed-loop control of the latter allows him to maintain a constant strain rate at the pole. This clearly must be verified for each new experimental arrangement and for each material.

An alternative technique, which seems to offer the possibility of obtaining results with less viscous polymers is based on the film blowing process. This is the analogue for biaxial stretching of spinning in uniaxial stretching, in the sense that it is a steady and generally inhomogeneous flow. Han and Park [H4] use the theory for a purely viscous liquid [P4] and calculate ratios of stress to rate of strain at points on the tubular film bubble. These "biaxial extensional viscosities" suffer from the same defect as "uniaxial extensional viscosities" obtained in spinning, since they are not material functions but depend on other experimental variables. Iwakura, Yoshinari and Fujimura [I7] attempt to simplify things by running the bubble with no stretching in the direction of flow (the "machine direction"). However their calculations of the stresses depend on having a bubble whose radius increases linearly with distance (a conical bubble), while they calculate (and observe experimentally) that the inverse square of the radius decreases linearly with distance. It is not clear how great an error this neglect of curvature in the machine direction introduces, and again there is the calculation of a viscosity on the assumption of purely viscous behaviour. There is therefore much that needs to be done in order to give a sound basis to film blowing as a rheometrical tool, and it looks at present distinctly less promising than spinning which, as we shall see (section 3.3) is by no means well founded. This is not to say, of course, that such experiments have no value, for they do give some information where we had none before. We cite further work on the analysis of the film blowing process below (section 7.2).

3.2.2 Experimental results for polymers

We list experimental work on biaxial extension in Table 3.2, including elastomers (not covered thoroughly) and thermoplastics. Work here is not so far advanced that much can be said about results except for polyisobutylene, which is a convenient material for these experiments. There seems to be general agreement that as the strain rate increases from very low values the viscosity decreases with strain rate for all types of deformation, and there is good evidence that viscosities in simple shear, uniaxial extension, strip biaxial extension and equal biaxial extension are in the ratio 1:3:4:6 even at strain rates where the material is no longer Newtonian but has a viscosity perhaps ten times lower than Newtonian [J13,P7,C26,D18,B2]. Two exceptions to this agreement should be mentioned; Vinogradov [V2] in uniaxial extension obtains a slight increase in η_T at strain rates above about $2 \times 10^{-2}5^{-1}$, and η_T is independent of $\dot{\gamma}$ below this strain rate, and Maerker and Schowalter [M9] found that η_{eb} goes through a minimum at around $4 \times 10^{-2} s^{-1}$ and then increases significantly. These results are both obtained at higher rates than any others mentioned here, and with materials of lower molecular mass. Hence it is impossible to state either that the upturn in the viscosity curve is a material property for some or all polyisobutylenes or that it is a result of uncertainties in the much more difficult experiments in question.

It would be most satisfying to the theorist if the predictions of the upper convected Maxwell model were in fact borne out qualitatively by these experiments, but such satisfaction should wait for further experimental encouragement.

Date	*Investigator*	*Polymer*	*Experimental technique*
THERMOPLASTICS			
1971	Denson & Gallo [D16]	Polyisobutylene	Bubble inflation
1972	Joyce, Poehlein & Denson [J12,J13]	Polyisobutylene	Bubble inflation
1973	Peng & Landel [P7]	Polyisobutylene	Flat sheet (strip biaxial)
1974	Maerker & Schowalter [M9]	Polyisobutylene	Bubble inflation
1974	Baily [B2]	Polyisobutylene	Bubble inflation
1974	Denson & Crady [D18]	Polyisobutylene	Bubble inflation (strip biaxial)
1975	Han & Park [H4]	High-density polyethylene Low-density polyethylene Polypropylene	Film blowing

Table 3.2 Experimental investigations of biaxial extension.

Date	Investigator	Polymer	Experimental technique
1975	Iwakura, Yoshinari & Fujimura [I7]	High-density polyethylene	Film blowing
1975	Lai & Holt [L2]	Polystyrene Polymethyl methacrylate	Bubble inflation (forming)
1975	Schmidt & Carley [S3]	Polystyrene Cellulose acetate butyrate	Bubble inflation
1976	Hoover & Tock [H12]	Polypropylene	Bubble inflation
1976	De Vries & Bonnebat [V12]	Polyvinyl chloride	Bubble inflation
1976	Denson & Hylton [D19]	ABS copolymer	Bubble inflation (strip biaxial)
1976	De Vries, Bonnebat & Beautemps [V13]	Polystyrene Polypropylene Polyethylene terephthalate	Bubble inflation
1978	Dealy & Rhi-Sausi [D8]	Low-density polyethylene	Bubble inflation
ELASTOMERS			
1944	Treloar [T22]	Natural rubber	Bubble inflation
1951	Rivlin & Saunders [R9]	Natural rubber	Flat sheet, bubble inflation
1964	Wessling & Alfrey [W9]	Natural rubber	Stretching over a conical mandrel
1966	Zapas [Z4]	Butyl rubber	
1968	Smith & Frederick [S12]	SBR	Tube extension & inflation (s.b.)
1970	McGuirt & Lianis [M3,M4]	SBR	Flat sheet (e.b., s.b.)
1970	Flowers & Lianis [F11]	SBR	Flat sheet
1971	Dickie & Smith [D21]	SBR	Bubble inflation
1975	Jones & Treloar [J8]	Natural rubber	Flat sheet
1977	Charrier & Li [C11]	Natural rubber	Tube extension & inflation

Table 3.2 (continued)

3.3 SPINNING AND RELATED FLOWS

3.3.1 Experimental techniques

We can usefully distinguish four major classes of experiment, namely nonisothermal melt spinning, isothermal melt spinning, solution spinning and the tubeless syphon or Fano flow. We use this classification in summarising experimental work in Table 3.3 below.

Nonisothermal melt spinning is basically the industrial process for the manufacture of artificial fibres of nylon, polyester and other polymers. Molton material is extruded downwards through a spinnerette (which is a small disc drilled with holes) or, in experimental work, through a die (a single hole), the thread is allowed to cool and solidify and then is taken up on a roll or drum. This is driven at a faster speed than the thread would attain under gravity and so the thread is stretched. Either the die or the take-up in an experimental arrangement will have a device for measuring the force exerted on one end or other of the thread being spun. The flow rate of polymer melt is measured and its temperature determined as a function of distance along the thread. This latter is not easy to do accurately. Also we require to measure either the velocity or the diameter of the thread as a function of distance. Commonly the latter is done by photographing the thread and measuring the size of the image but an alternative is to trap and weigh predetermined lengths of the thread [K9]. The measurement of diameter needs great care, especially since the data will have to be differentiated to obtain the strain rate as a function of distance along the thread. A choice here is whether to smooth the data, for example by fitting a polynomial or a set of splines, or not.

The other piece of information required is the stress acting on any cross-section of the thread. We obtain this by considering the axial components of forces acting on the portion of thread between an arbitrary cross-section at X and the end of the thread where the force is applied ($X = L$) and since we are considering a real experiment on a real material (in contrast to the idealisation of the previous chapter) we cannot ignore all additional forces. The net downward force on the portion of thread is the applied force, F, plus the weight of that portion of the thread minus the air drag, the net surface tension force and the stress at cross-section T_X multiplied by the area, $A(X)$, and this must give the rate of increase of momentum of the melt. This is conveniently written (following Ziabicki [Z8,Z9])

$$T_X A(X) = F + F_{gravity} - F_{drag} - F_{surface} - F_{inertia} \qquad (3.1)$$

where

$$F_{gravity} = \int_X^L \rho g A(x)\,dx \qquad (3.2a)$$

for downward spinning into air (negligible buoyancy effect, $\rho >> \rho_{air}$)

$$F_{drag} = \int_X^L \tfrac{1}{2}\rho_{air} U(x)^2 C_f S(x)\,dx \qquad (3.2b)$$

where C_f is the skin friction coefficient (not a constant) and $S(x)$ is the perimeter of the cross-section at x,

$$F_{surface} = \sigma(S(X) - S(L)) \qquad (3.2c)$$

where σ is the surface tension coefficient between the air and the melt, and

$$F_{inertia} = \rho Q(U(L) - U(X)) \qquad (3.2d)$$

is the inertia force representing the net rate of increase of momentum of the melt.

A little care has to be taken in dealing with the surface tension contribution when we are dealing with an incompressible material. In this case T_X is not determined by the constitutive equation, but is given by

$$T_X = p_{11} - p \qquad (3.3)$$

as was used in (2.66) above. Then when surface tension is taken into consideration (2.78) is replaced by

$$p_{22} - p = -\sigma/R \qquad (3.4)$$

in the approximation where R' and R'' are small compared with R. Hence when we eliminate the isotropic pressure p we find

$$T_X = p_{11} - p_{22} - \sigma/R \qquad (3.5)$$

and so

$$(p_{11} - p_{22})A(X) = T_X A(X) + \pi\sigma R(X)\ . \qquad (3.6)$$

Then (provided we are dealing with a thread with a circular cross-section) we may replace T_X by $p_{11}-p_{22} = T$ in (3.1) if we replace (3.2c) by half the value given there, namely

$$F_{surface} = \sigma(\pi R(X) - \pi R(L)) \, . \tag{3.7}$$

Generally it is found that surface tension has a negligible influence in melt spinning. If we are lucky inertia will also be negligible (though we cannot make this approximation for solutions) and gravity will more or less balance air drag [W13]. Hence it is often possible to equate the measured force to the force at any cross-section. This simplification can be used in an approximate treatment of data. However it is not recommended that this approach be used for work with accurate data, and air drag, gravity and inertia (in order of probable decreasing importance for melts) may between them contribute as much as 50% of the applied force in practical situations [Z9,B7].

Isothermal melt spinning experiments have been carried out by a number of workers. The basic technique is to surround the first part of the thread with an "isothermal chamber", and to use measurements of thread diameter in this region where the melt is maintained at a constant temperature. The force measurement is made and treated in the same way as in the non-isothermal case. One variation due to Fehn [F2] is the use of tracer particles to measure velocity directly (from frame by frame analysis of cine film) instead of using diameter measurements. This appears to require much more work, but may offer greater accuracy; a proper analysis of experimental errors involved in both techniques would be a useful exercise (see also [W8] p.474). The value of isothermal experiments in separating the basic rheological response of the material from the temperature-dependence of its behaviour should be obvious for any fundamental study of polymer melts and, coupled with heat transfer studies, offers a more reliable understanding of the industrial process, even though temperature changes may exert a dominant influence on the flow.

Spinning experiments have also been carried out for solutions and suspensions, and the basic arrangement is very similar to that for melts [W8, M29,F4]. The experiments are now naturally isothermal (though of course precautions need to be taken to regulate the temperature of the liquid and of the surroundings). The take-up is usually on a wheel to which the liquid adheres, and which is driven at selected speed. As was remarked above, we can no longer, even in a first approximation, ignore gravity or inertia though air drag is not usually important since filament velocities are a good deal smaller than for commercial melt spinning and while surface tension will be more important than it is for melts, it too may not be too important. Moore and Pearson [M33] show graphically the relative importance of all these factors for two sets of experimental conditions. We include in the discussion of spinning the essentially similar flow generated by the triple jet apparatus of Oliver [O4]. The only fundamental difference is that the stretching force is not supplied by a rotating wheel, but by three small high velocity jets impinging on the thread or main jet. This apparatus involves a jet shorter than the thread in most spinning experiments, and higher velocities and strain rates are obtained.

Finally there are a number of experiments using the tubeless syphon (see frontispiece), a flow whose discovery is attributed to Fano [F1]. In this arrangement the liquid is sucked up through a tube whose end is raised above the surface of the reservoir of liquid. The flow from the reservoir to the tube is kinematically the same as that obtained in spinning upwards but is possible for fewer liquids (depending on the combination of factors influencing 'spinnability'). The major difference, apart from the direction of the gravity force, is in the initial conditions, which are as uncertain as those for spinning, but do not involve a prior history of shearing or the emergence of the liquid from an orifice. The latest version of the Fano flow, due to Balmer and Hochschild [B6], employs a pressurised chamber so that greater flow rates can be obtained, and they succeed in obtaining 2mm columns of water by this technique. There is also a report [W4], with limited quantitative data, of a similar flow for molten polyethylene terephthalate. This is non-isothermal and intended for fibre manufacture, but is very limited in its operation. Instead of suction the flow is driven by winding up the solidified fibre, which draws the molten polymer from a pool.

3.3.2 Experimental results

Table 3.3 records rheological experiments on spinning and related flows according to the classification discussed above. In addition melt spinning experiments where the thread shape (and perhaps temperature) have been recorded but not the force, and where the force is recorded but not the shape, are noted. These have clear technical value and are also of some use to the rheologist. In particular the 'melt strength' type of experiment shows marked differences in behaviour between different polymers, notably between branched or low density polyethylene (LDPE) and linear or high density polyethylene (HDPE) [B30]. The test is normally carried out by measuring the force in a melt spinning experiment while the take-up velocity is steadily increased until the thread breaks. Both the force, which becomes more or less independent of take-up speed except at the very lowest speeds, and the maximum take-up speed, are useful practical indications of how polymers will behave in forming and shaping flows with significant elongational components. The observed constancy of the force, and the markedly lower force and higher maximum take-up speed of HDPE, compared with LDPE, are phenomena which the rheologist and the applied mathematician might seek to explain.

The non-isothermal spinning experiments must also be classed as being more of technical than of fundamental importance, and results of a number of workers illustrate this. It has been found [K9,K10,S2,I5,L12] that if the ratio of stress to strain rate at a point on the thread is calculated (usually this is called an apparent elongational viscosity - a potentially misleading terminology we discuss further below - then this may be correlated satisfactorily with the temperature at that point, and no dependence on strain rate or strain history is apparently necessary to describe the data. The results of Sano, Orii and Yamada [S2] for polypropylene go even further and show that, to within an order of magnitude at least the ratio of stress to strain rate at a point three times the (Newtonian) shear viscosity for the temperature at the point. (This work is reported in some detail by Ziabicki [Z9].)

There seems to be no doubt that the influence of temperature on the rheology is the dominant influence, but the evidence is not really adequate to say more than that. Recent work by Matsui and Bogue on HDPE [M18] provides more detailed evidence, from analysis of data of Dees [D9]. They give theoretical predictions of stress based both on a viscous theory (stress is strain rate multiplied by three times the shear viscosity at that appropriate temperature) and a viscoelastic theory (a non-isothermal form of the Bogue-White model). Both of these give adequate stress predictions over the latter two-thirds of the thread, but the viscoelastic theory is much better over the first third, where the influence of the previous shearing flow is apparently still important. (A number of the earlier experiments do not report data obtained in the early part of the spinline.)

The influence of spinning conditions is demonstrated by Cernia and Conti [C4] who plot the ratio of local stress to strain rate again local temperature and obtain two distinct curves for two different final thread diameters. The difference is less than the order of magnitude scatter of Sano, Orii and Yamada [S2] but Cernia and Conti's data, on one particular polyethylene terephthalate (PET), shows less scatter and the distinction is quite definite. They argue further that the results cannot be accepted as viscosity - temperature data because the temperature dependence (expressed as an activation energy) is different for the different final diameters and also because the activation energy is much lower than that for shear flow. Cernia and Conti seek to describe their results in terms of a simple visco-elastic model which we discuss below. An easier approach is the empirical one of Hill and Cuculo [H7], who seek to describe their PET data by a linear dependence of the logarithm of the ratio of stress to strain rate on the reciprocal of temperature, the strain rate, and the product of these two variables. This gives an activation energy which is a decreasing function of strain rate, and also confirms that temperature is undoubtedly the dominant factor in determining the stress in the thread.

The problems we have been discussing in the interpretation of data serve to highlight the need for isothermal experiments in order to uncouple rheological and thermal effects (though of course we must understand both, and their interaction in any full understanding of practical melt spinning). There are two main points to be made in connection with isothermal spinning data for melts and solutions. First the practice of prescribing the ratio of stress to strain rate at a point on the spinline as an "elongational viscosity" is to be avoided, and even the description of the ratio as an "apparent elongational viscosity" seems open to misinterpretation. The point is well illustrated by data of Deprez and Bontinck [D20] who show data for HDPE with two distinct curves for different values of the initial velocity of the thread. The same sort of data, emphasising the failure of an interpretation based on an "apparent elongational viscosity" which depends on the local strain rate, is to be found in other papers [C14,S15,B7]. The practical point is that any such correlation offers no guarantee that it may be used successfully in scaling or in any attempt at predicting behaviour where parameters such as flow rate or thread diameter are altered appreciably.

The second point concerns the idea employed in some early studies [A1,C14] of attempting to allow for initial conditions (in particular the axial stress at the die exit) by subtracting a term representing the decay of the initial stress at a rate given by an average (or even a maximum) relaxation time. This can be shown to be a wrong assumption about the rate of decay for simple theoretical models [P18], but more important is the fact that the stress we have discussed calculating in the previous section is the stress at a cross-section of the thread, and the attempt to split this into a decaying initial stress which does not stretch the material and a residual stress which does the stretching is implicitly postulating a rheological model of a very unusual kind, which has not been investigated in any other context. Of course it is natural to wish to calculate the response of virgin material but we shall never find this in a spinning threadline. It may be possible to take account of the initial stress and the influence of the flow history in some way, but this seems feasible only in the context of specific rheological models. It must be wrong to think of the initial stress as something to exclude from force balances on the thread, just as it is wrong to postulate that the initial stress at the point of maximum diameter is zero. (This fallacy is mentioned by Ziabicki [Z9] pp. 177-8, in discussing early work on modelling the melt spinning process.) The above treatment of initial stress is also to be found in work on spinning polymer solutions [F4,H13,H14,M5] where it is used to reduce 'apparent elongational viscosity' data to a single curve - an empirical success which the author views as a case of two 'wrongs' making not a 'right' but a tidy graph. (See also section 6.2.1)

Another successful idea for correlating data is that of relating stress to strain rather than strain rate. This appears to have been proposed first by Fehn [F2] for HDPE, but more recently has found more popularity for polymer solution spinning results [H14,O5,O6]. The danger here is the old one of ending up plotting x^2 against x rather than plotting two independent quantities. For example, if gravity, air drag and surface tension may be neglected (leaving inertia and the applied force) we have the stress

$$T = (T_o/V_o - pV_o)\, V + pV^2 \qquad (3.8)$$

(equation (4) of [O5]) and then defining the strain

$$\gamma = \text{Log}\,(V/V_o) \tag{3.9}$$

gives the relation

$$T = Ae^{\gamma} + Be^{2\gamma} \tag{3.10}$$

where $A = T_o - pV_o^2$ and $B = pV_o^2$, which will vary between different experiments (as in the results of Oliver [O5], figures 3 and 4). There is then no rheological information in this equation (or the corresponding graphs) but merely a record of the way the stress has been calculated. In fact, if a material were purely elastic, in the sense that T was a function of γ only then either the constitutive equation is the same as (3.10) and we do not have a uniquely determined threadline shape (no information is obtained on the X dependence of T and V or γ) or else the constitutive equation is different from (3.10) and inconsistent with it almost everywhere, so that there is no threadline shape which will satisfy both equations. Of course it is quite possible to have a constitutive equation of the form suggested by Waters, King and Oliver [W6], where the stress depends on strain and strain rate. Use of this together with (3.10) gives a differential equation for the strain, with the property that the strain becomes constant in agreement with some experimental observations (of constant threadline diameter). See also the discussion in section 5.3.2 of the asymptotic form of equations for stretching at large strain rate.

An observation of threadline shape which is very convenient for analysis may be noted, namely that of Spearot and Metzner [S15] for LDPE and of Acierno and co-workers [A2,A3] for dilute and concentrated polyacrylamide solutions, who find that the strain rate often does not vary significantly along the threadline. Kanel [K3] has generalised this and gave the term Km Kinematics to threadline flows which satisfy

$$D^{-m} = AX + B \tag{3.11}$$

for same m, with constant A and B. (D is the diameter of the thread at position X, and m = 2 corresponds to a constant strain rate.) Data with m from 0.6 up to 2.7 are reported in Kanel's thesis, mainly his own data and that of Weinberger [W8]. It is not clear how helpful this idea is, or whether the related idea, due to Weinberger and Goddard, that spinning flows form a "monogenic family of extensions" can usefully be sustained. We shall return to this idea later, but note here that the bulk of the evidence both experimental and theoretical does not seem to favour the idea as a general one since it implies that a liquid would have a unique "spinning viscosity". In fact different workers, either using slightly different polymers or spinning conditions obtain qualitatively different results, for example Zidan [Z10] has a strain rate which decreases with increasing distance down the spinline for nominally the same material as used by Weinberger and by Acierno et al. who find, in most of their experiments, an increasing and a constant strain rate. The same may be seen in the different results of Han and Lamonte [H2] and Deprez and Bontinck [D20] for HDPE; not a disagreement but an indication that spinning conditions have a very great effect on the kinematics. Systematic explorations of these effects are to be found, for example, in the work of Deprez and Bontinck who found that for LDPE the strain rate is more or less constant at low draw ratios (e.g. 4), but increases along the spinline for high draw ratios (e.g. 16) [D20] and similar results are reported by Moore and Pearson [M33] for dilute polyacrylamide solutions.

One final point must be made, concerning data for polymer solutions. This concerns the very great difference between the behaviour of dilute and concentrated solutions discussed by Acierno, Titomanlio and Greco [A2]. The stresses developed in dilute solutions in elongational flows are much higher, relative to the stresses in shear flow, than for concentrated solutions. If we think (for a moment) in terms of "apparent elongational viscosity" and define the Trouton ratio as the ratio of this quantity to the shear viscosity at zero shear rate then for dilute solution this ratio is typically from 200 to 2000, while for concentrated solutions it takes values from 3 to 20, and approaches 3 at low strain rates. In this respect polymer melts behave like concentrated solutions, and concentrated can mean, for example, a 2% (by weight) aqueous solution of polyacrylamide (a $\frac{1}{4}$% of this polymer is dilute). The distinction between dilute and concentrated is made on the basis of whether entanglements between molecules are a significant feature affecting the dynamics, and in practice we may make this distinction at the concentration where the viscosity increase with concentration becomes markedly nonlinear (see for example Bird, Armstrong and Hassager [B18] pp. 76-78). We discuss below (section 4.1) attempts to explain this behaviour at a molecular level.

Date	*Investigator*	*Polymer*
ISOTHERMAL MELT SPINNING (FORCE AND FILAMENT SHAPE MEASURED)		
1968	Fehn [F2]	HDPE
1971	Acierno, Dalton, Rodriguez & White [A1]	LDPE, PS
1972	Chen, Hagler, Abbot, Bogue & White [C14]	LDPE, HDPE
1972	Han & Lamonte [H2]	LDPE, HDPE, PP, PS
1972	Spearot & Metzner [S15]	LDPE
1973	Zeichner [Z5]	PS
1974	Fehn [F3]	PVC
1974	Han & Kim [H3]	PS/HDPE blend, filled PS
1975	Deprez & Bontinck [D20]	LDPE, HDPE
1977	Bankar, Spruiell & White [B7]	Nylon-6
1978	Bayer, Schreiner & Ruland [B9]	LDPE
NONISOTHERMAL MELT SPINNING (FORCE AND FILAMENT SHAPE MEASURED)		
1961	Ziabicki [Z8]	Polyamide
1965	Kase & Matsuo [K9]	PET
1967	Kase & Matsuo [K10]	PP
1968	Sano, Orii & Yamada [S2]	PP
1970	Ishibashi, Aoki & Ishii [I5]	Nylon-6
1973	Cernia & Conti [C4]	PET
1974	Hill & Cuculo [H7]	PET
1974	Lin & Hauenstein [L12]	PET
1976	Matsui & Bogue [M18]	HDPE
1977	Banker, Spruiell & White [B7]	Nylon-6
1977	Blyler & Gieniewski [B22]	Styrene-silicone copolymer

PS = polystyrene, PVC = polyvinyl chloride, PET = polyethylene terephthalate, PP = polypropylene.

Table 3.3 *Spinning and related experiments.*

Date	*Investigator*	*Polymer*
MELT SPINNING FILAMENT SHAPES (NO FORCE)		
1959	Ziabicki & Kedzierska [Z7]	Polymide, polyester
1966	Wilhelm [W20]	Polyester
1967	Copley & Chamberlain [C36]	Nylon-6
1971	Kohler [K16]	PET
1974	Dees & Spruiell [D9]	PP
MELT SPINNING FORCE MEASUREMENTS (NO PROFILES)		
1966	Bergonzoni & di Cresce [B10]	PP, LDPE, HDPE, PS
1967	Busse [B30]	LDPE, HDPE, ionomers
1967	Kaltenbacher, Howard & Parson [K1]	LDPE
1969	Cogswell [C25]	LDPE
1971	Meissner [M20]	LDPE, HDPE
1972	Meissner [M21,M23]	LDPE
1973	Wissbrun [W24]	HDPE, PP
1975	Deprez & Bontinck [D20]	LDPE, HDPE
TUBELESS SYHPON (FANO FLOW)		
1970	Astarita & Nicodemo [A16]	½% PAA
1972	Kanel [K3]	Dilute PAA
1974	Acierno, Titomanlio & Nicodemo [A3]	Dilute PAA
1975	Nicodemo, de Cindio & Nicolais [N5]	Dilute PAA*
1976	de Cindio, Nicodemo, Nicolais & Ranaudo [C17]	Dilute PAA*
1977	Peng & Landel [P8]	Dilute polymer solutions

H(L)DPE = High-(low-) density polyethylene, * Suspension of glass beads

Table 3.3 (*continued*)

Date	*Investigator*	*Polymer*
SPINNING FOR POLYMER SOLUTIONS AND SUSPENSIONS		
1969	Zidan [Z10]	3% PAA, Viscose, Newtonian oil
1974	Acierno, Titomanlio & Greco [A2]	Conc: PAA (2%)
1974	Acierno, Titomanlio & Nicodemo [A3]	Dilute PAA ($\frac{1}{4}$%)
1974	Hudson, Ferguson & Mackie [H13]	Conc: polybutadiene
1974	Mewis & Metzner [M29]	Fibre suspension (Newtonian oil)
1974	Oliver & Bragg [O4] (cf. [O5,O6]	Dilute PAA, Polyox
1974	Weinberger & Goddard [W8]	1$\frac{1}{2}$% PAA, Polyox, Fibre suspension (Newtonian oil)
1975	Moore & Pearson [M33]	1% PAA
1976	Hudson & Ferguson [H14]	Conc: polyacrylonitrile
1977	Oliver & Ashton [O7]	Dilute PAA, Polyox
1977	Baid & Metzner [B1]	Dilute PAA
1978	McKay, Ferguson & Hudson [M5]	Conc: polyurethane
1978	Chang & Denn [C9]	Corn syrup, Dilute PAA

PAA = polyacrylamide, Polyox = polyethylene oxide

Table 3.3 *(continued)*

4 Flow classification

"The incredulous considered it, a priori, an insipid and laborious theological game."

Jorge Luis Borges.

We have managed to avoid a formal definition of an elongational flow so far, and in fact have given the reader no means of determining whether a flow other than one of our basic flows is elongational. Here we shall attempt to remedy this omission and at the same time discuss a number of ways of classifying flows.

The contrast between behaviour of polymeric liquids in shear and in elongation leads us to wish to generalise these ideas. The resultant classification into weak and strong flows is motivated by ideas of what happens to long chain macromolecules. We shall illustrate some ideas by a more detailed consideration of two-dimensional flows. An attempt to define elongational flows in a general way will necessitate an excursion into formal continuum mechanics and the decomposition of deformation into pure stretching and rotation. We shall confine our attention for the most part to steady and spatially homogeneous flows.

4.1 WEAK AND STRONG FLOWS

4.1.1 Stretching macromolecules

Early discussions of molecular behaviour in the flow of dilute polymer solutions deal with simple shear [P10], and Ziabicki [Z6] appears to be the first person to make the distinction between shear flows (flows with transverse velocity gradient in his terminology) and elongational flows (flows with parallel velocity gradient). Detailed investigation of some molecular models by Takserman-Krozer [T2] illustrated the differences between shear and elongation, and led (among other things) to the conclusion that phenomena in elongation "cannot be explained in terms of experiments or theories concerning shear flow". Giesekus [G8] reaches a similar conclusion, following a detailed analysis of possible velocity fields with constant velocity gradient [G6]. There is an extremely clear statement of the essential points of difference between shear and elongation by Frank [F13]. The most important of these is that in shear if a molecule gets extended it is oriented parallel to the streamlines and lies more or less in fluid of uniform velocity (because the velocity does not change along the streamlines) and does not affect the motion greatly, while in elongation (where the velocity does change along the streamlines) the ends of a molecule are in fluid of increasingly different velocity so that once the velocity gradient is large enough to overcome the natural tendency of the molecule to coil up there is a very rapid change to the extended form which affects the motion significantly. De Gennes [G4], Tanner [T11] and Hinch [H8] discuss these ideas in more detail, and show how a hysteresis effect may be expected, associated with a range of elongation rates where both the stretched and coiled molecular arrangements are possible (and stable to small disturbances).

The enormous effect of long slender particles on elongational viscosity has been demonstrated experimentally by Mewis and Metzner [M29] and by Weinberger and Goddard [W8] and explained theoretically by Batchelor [B8]. We may suppose that the smaller difference between shearing and elongation for concentrated solutions and melts is due to entanglements between molecules. These may on the one hand hinder the tendency of the molecules to be lined up in elongation, and on the other hand to increase the interaction of the molecules with a shear flow. Both of these can be thought of in terms of a reduced effective aspect ratio of a molecule.

It may be helpful to examine the differences between shear and elongation in a little more detail. In simple shear one set of parallel material planes slide over each other, and every material line not in one of these planes rotates in the same direction. We can think of molecules in the flow being rotated and stretched until they are lined up in the shearing planes and parallel to the flow. In this situation any disturbance (such as Brownian motion) causing rotation in the same direction will lead to a molecule having to rotate by a full 180° before it is lined up again, so we must expect to find some molecules not lined up however long the flow persists. The number is not, on average, very large according to estimates of the time taken for the rotation. A much more important fact is that a molecule which is lined up experiences a stretching force depending on the shear rate and the width of the molecule (the product of these determines the difference in fluid velocity at the extremities of the molecule).

In uniaxial elongation material lines in the direction of elongation are stretched, and those perpendicular to this direction are shortened (by half as much, per unit length, if the material is incompressible). All other material lines rotate towards the direction of elongation by the shortest route, so there is rotation of material lines about all axes perpendicular to the direction of elongation, and the average rotation of a material element is zero (the vorticity is zero, in contrast to simple shear where the vorticity has the same magnitude as the rate of strain). Thus polymer molecules are again lined up and stretched in the direction of flow, but with two important differences. If a molecule is disturbed from this direction the hydrodynamic restoring force is directed straight back to the equilibrium position so that there is no tumbling motion as in shear, and once the flow has lasted for long enough all the molecules will be exactly or very nearly lined up. Also the two ends of a lined up molecule experience a velocity difference proportional to the length of the molecule and the rate of strain (or velocity gradient).

4.1.2 A kinematic classification

We are led to the idea of classifying flows into elongation-like and shear-like on the basis of molecular behaviour. The questions which arise for a particular flow are firstly a purely kinematic one - could the flow be responsible for extending polymer molecules significantly? - and secondly a more realistic or practical question - will the flow cause molecules of polymer in a given liquid to be extended significantly? The important quantities in this discussion are the eigenvalues of the velocity gradient tensor (which have the dimensions of $(\text{time})^{-1}$) and also, for the second question, a characteristic relaxation time for a polymer molecule in the liquid.

Tanner and Huilgol [T12] make the important purely kinematic classification in a simplification of Giesekus' work [G6], and define a flow as "strong" if the matrix of Cartesian components,

$$L_{ij} = \partial U_i / \partial X_j \,, \qquad (4.1)$$

of the velocity gradient tensor $\underline{L}$ has any eigenvalues with positive real part. In a homogeneous flow two material particles with relative position vector $\underline{r}$ separate at a rate

$$\dot{\underline{r}} = \underline{L} \cdot \underline{r} \qquad (4.2)$$

and for $\underline{L}$ independent of time this has the solution

$$\underline{r} = \underline{r}_o \exp(\underline{L}t) \tag{4.3}$$

which grows exponentially with time if any eigenvalue of $\underline{L}$ has a positive real part. Now for an incompressible fluid tr($\underline{L}$) is zero, so the sum of the eigenvalues of $\underline{L}$ is zero and hence a flow will be strong unless all the eigenvalues are zero or purely imaginary.

Now if all the eigenvalues are zero the flow is either a viscometric flow or an MCSH-II (section 4.3.5), and the converse is also true. This assertion is proved [T12] by considering the Jordan normal form of $\underline{L}$ when $\underline{L}$ has zero eigenvalues. We deduce that if

$$\underline{L} = \underline{P} \cdot \begin{pmatrix} 0 & 1 & 0 \\ 0 & 0 & 0 \\ 0 & 0 & 0 \end{pmatrix} \cdot \underline{P}^{-1} \tag{4.4a}$$

or

$$\underline{L} = \underline{P} \cdot \begin{pmatrix} 0 & 1 & 0 \\ 0 & 0 & 1 \\ 0 & 0 & 0 \end{pmatrix} \cdot \underline{P}^{-1} \tag{4.4b}$$

then $\underline{L}^2$ or $\underline{L}^3$ respectively are zero. For the converse it is simplest to use the basic definition of the eigenvalue

$$\underline{L} \cdot \underline{y} = \lambda \underline{y} \tag{4.5}$$

for some non-zero $\underline{y}$. By repeated use of this definition we obtain

$$\underline{L}^n \cdot \underline{y} = \lambda^n \underline{y} \tag{4.6}$$

and hence if $\underline{L}^n = \underline{0}$ we must have $\lambda = 0$ for any eigenvalue. Hence if a flow is viscometric or an MCSH-II it is weak. An MCSH-III (section 4.3.5) may be weak only if its three eigenvalues are distinct and two are purely imaginary (and complex conjugates) with the third being zero.

Thus weak flows include orthogonal rheometer flow [T12] as well as MCSH-II and viscometric flows. Hinch [H8] remarks that such flows "are very rare in the class of all possible types of flow", and Mackley [M7] makes a similar observation. Strong flows clearly are all flows where $\underline{L}$ has any real non-zero eigenvalue (either having three real distinct eigenvalues whose sum is zero, real eigenvalues $\alpha,\alpha,-2\alpha$, or complex eigenvalues $\alpha + i\beta$, $\alpha - i\beta$, -2α). Giesekus [G6] and De Gennes [G4] both discuss conditions for the various types of eigenvalue of $\underline{L}$ in more detail, obtaining these in terms of the invariants of $\underline{e}$ and $\underline{\omega}$. This is done by choosing for coordinate axes the principal axes of $\underline{e}$, so that

$$\underline{L} = \underline{e} - \underline{\omega} = \begin{pmatrix} e_1 & -\frac{1}{2}\omega_3 & \frac{1}{2}\omega_2 \\ \frac{1}{2}\omega_3 & e_2 & -\frac{1}{2}\omega_1 \\ -\frac{1}{2}\omega_2 & \frac{1}{2}\omega_1 & e_3 \end{pmatrix} \tag{4.7}$$

We have written e_i for the eigenvalues of $\underline{e}$, and ω_i for the components of the vorticity vector

$$\nabla \times \underline{U} \equiv \text{curl}(\underline{U}) = (\omega_1, \omega_2, \omega_3) \tag{4.8a}$$

which is related to the vorticity tensor (2.14) by

$$\omega_{ij} = \tfrac{1}{2} \sum_{k=1}^{3} \varepsilon_{ijk}\, \omega_k \,. \tag{4.8b}$$

The critical cases [G4] where the type and sign of the eigenvalues of $\underline{L}$ may change correspond either to a repeated root of the eigenvalue equation or to a zero eigenvalue. The eigenvalue equation

$$\det(\underline{L} - \alpha\underline{I}) = 0 \tag{4.9a}$$

may be written explicitly for (4.7) as

$$\alpha^3 + P\alpha + Q = 0 \tag{4.9b}$$

where

$$P = \tfrac{1}{4} \sum_{i=1}^{3} \omega_i^2 - \tfrac{1}{2} \sum_{i=1}^{3} e_i^2 \tag{4.9c}$$

$$Q = -e_1 e_2 e_3 - \tfrac{1}{4} \sum_{i=1}^{3} e_i\, \omega_i^2 \,. \tag{4.9d}$$

Then we may obtain the critical conditions

$$27Q^2 + 4P^3 = 0 \tag{4.10}$$

(with $P < 0$ and $Q > 0$) for a repeated root,

and

$$Q = 0 \tag{4.11a}$$

for a zero root, with the additional condition

$$P > 0 \tag{4.11b}$$

so that the other roots are purely imaginary.

The latter condition does correspond to transition between weak and strong flows as we have defined them above, but the former, (4.10), may correspond also to a transition between strong flows with all real eigenvalues and with a pair of complex eigenvalues. The condition $Q > 0$ means that the third root is negative, so De Gennes appears to be saying that a flow for which L has complex eigenvalues with positive real part will not stretch the molecules. However his argument is based on the assumption that a long molecule in a steady velocity field will take up a fixed orientation. This will certainly happen in an elongational flow, but in the flow with eigenvalues $(\alpha + i\beta, \alpha - i\beta, -2\alpha)$ for $\alpha > 0$ we may expect a long molecule with one end at the origin to whirl around the X_3-axis while being stretched along its length in a (rotating) direction perpendicular to the X_3-axis. Hence it seems that Tanner and Huilgol's classification is the correct one.

4.1.3 Dynamics of a macromolecule

So far in this discussion we have been concerned solely with the separation of fluid particles. The stretching of polymer molecules by the fluid is also influenced by the natural tendency of a long flexible molecule to coil up. De Gennes [G4] avoids this by postulating that the velocity gradients are very large, and hence he is concerned to show the effect of different types of flow rather than of the magnitude of the velocity gradient. His results show that there are some flows where a molecule will naturally assume a coiled configuration (no matter how great the velocity gradient) and others where (provided the velocity gradient is large enough) the molecule will be stretched out. In order to clarify this latter point and discover how large a velocity gradient we require, it is necessary to consider the forces acting on a long molecule.

For simplicity we follow Tanner [T13] and consider an elastic dumbbell, with a spring, which represents the natural tendency of a molecule to coil up, connecting two beads, on which the fluid drag acts. If we have a linear spring with force $K\underline{r}$ and Stoke's law drag $\zeta(\underline{U}-\dot{\underline{r}})$ we find

$$\dot{\underline{r}} = \underline{L} \cdot \underline{r} - (2K/\zeta)\,\underline{r} \tag{4.12}$$

Here $\underline{r}$ is the position vector of one bead relative to the other and $\zeta/2K$ is a characteristic relaxation time for the elastic dumbbell. There is some variation in usage here, between those who use this characteristic time [G4,F13,H8,M7] and those who use half this value [T13,D14,M13]. The reason for the latter choice is that the evolution equation for the average $\langle \underline{r}\,\underline{r} \rangle$ in the statistical mechanics of suspensions of macromolecules is (neglecting Brownian motion) [B19]

$$\langle \underline{r}\dot{}\,\underline{r} \rangle = \underline{L}\cdot\langle \underline{r}\,\underline{r} \rangle + \langle \underline{r}\,\underline{r} \rangle\cdot\underline{L}^{T} - (4K/\zeta)\langle \underline{r}\,\underline{r} \rangle \qquad (4.13)$$

and it is from this that the simplest theory gives the upper convected Maxwell model with $\lambda = \zeta/4K$. We shall adopt the latter convention which is more natural in continuum mechanics (just as the former is more natural in considering the physics at a molecular level).

The important quantities for the elastic dumbbell are the eigenvalues of $\underline{L} - (1/2\lambda)\,\underline{I}$, and our earlier discussion relates to the case where $|2\lambda\alpha_i| >> 1$ for the eigenvalues α_i of $\underline{L}$. By using the Routh-Hurwitz criterion Tanner [T13] obtains the conditions

$$\det(\lambda\underline{L} - \tfrac{1}{2}\underline{I}) < 0 \qquad (4.14a)$$

and

$$\det(\lambda\underline{L} - \tfrac{1}{2}\underline{I}) > \tfrac{3}{4}\mathrm{tr}((\lambda\underline{L} - \tfrac{1}{2}\underline{I})^{2}) - 27/16 \qquad (4.14b)$$

for a flow to be weak. If either or both of these is not satisfied then the flow is "strong". "Strong" has now got a slightly altered meaning, no longer purely kinematic, and a steady flow is "strong" if it will fully extend a molecule with relaxation time $\frac{1}{2}\lambda$. The linear spring is infinitely extended in such a case, corresponding to the infinite stresses required for steady flow of the upper convected Maxwell model in elongational flows above the critical strain rate. This is generalised to the notion of critical kinematic states [T10] and is one idea pointing to the need for more realistic finitely extensible dumbbell models [B19,T9].

For more realistic models of long chain polymers another feature of importance is the influence of molecular mass on relaxation time and hence on critical strain rates. Mackley and Keller [M6] suggest that a representative relaxation time may be taken to vary with molecular mass to the power 7/4, and hence find a critical molecular mass

$$M_c \propto (\dot{\gamma})^{-(4/7)} \tag{4.15}$$

above which a molecule will be fully extended in a flow with elongation rate $\dot{\gamma}$. For their solutions of polyethylene in xylene they estimate $M_c \simeq 10^8$ for $\dot{\gamma} = 1s^{-1}$ and $M_c \simeq 2 \times 10^6$ for $\dot{\gamma} = 1000s^{-1}$. This leads to the conclusion that with the high rate of strain a small but significant part of a polymer of average molecular mass 10^5 would be fully extended. The importance of the high molecular mass tail in a molecular mass distribution is thus emphasised. It is also to be remarked [T9] that breadth of molecular mass distribution smooths out the suddenness of the transition from coiled to stretched molecules in a solution.

One final important aspect of the stretching of long chain molecules by strong flows is the question of whether a molecule stays in the flow for long enough to be stretched. We shall not pursue this here, but refer the reader to the discussion of the Deborah number and some applications by Tanner [T13]. Marrucci [M13] and Mackley [M7] also discuss this point. While most of this discussion has been centred around dilute solution ideas (where separate molecules are independent) similar considerations will apply to concentrated solutions and melts, with the main difference that in the highly entangled state the difference between weak and strong flows is much less marked.

4.2 TWO-DIMENSIONAL FLOWS

4.2.1 Steady homogeneous plane flows

It is possible, as Giesekus has done [G6], to investigate general three-dimensional flows in terms of the rate of strain and vorticity tensors, but we shall do this only for the simpler special case of two-dimensional flows. These have been investigated by Marrucci and Astarita [M10] as well as by Giesekus [G7]. Given the constraint of isochoric flow (for an incompressible fluid) the general homogeneous steady flow in two dimensions has the velocity gradient

$$\underline{\hat{L}} = \begin{pmatrix} \dot{\gamma} & \hat{\kappa}_1 \\ \hat{\kappa}_2 & -\dot{\gamma} \end{pmatrix} . \tag{4.16}$$

This may be reduced to the canonical form

$$\underline{L} = \begin{pmatrix} 0 & \kappa_1 \\ \kappa_2 & 0 \end{pmatrix} \tag{4.17}$$

where

$$\kappa_1 = \tfrac{1}{2}(\hat{\kappa}_1 - \hat{\kappa}_2) + \surd(\dot{\gamma}^2 + \hat{\kappa}^2) \tag{4.17a}$$

$$\kappa_2 = \tfrac{1}{2}(\hat{\kappa}_2 - \hat{\kappa}_1) + \surd(\dot{\gamma}^2 + \hat{\kappa}^2) \tag{4.17b}$$

$$\hat{\kappa} = \tfrac{1}{2}(\hat{\kappa}_1 + \hat{\kappa}_2) \tag{4.17c}$$

by rotating the axes through an angle θ given by

$$\sin 2\theta = \hat{\kappa}/\surd(\dot{\gamma}^2 + \hat{\kappa}^2). \tag{4.17d}$$

Simple shear ($\kappa_2 = 0$) and pure shear (or strip biaxial extension; $\kappa_2 = \kappa_1$) are clearly special cases of this. The eigenvalues of $\underline{L}$ are $\pm\surd(\kappa_1\kappa_2)$ and so we have a weak flow if $\kappa_1\kappa_2 < 0$, and we then have a flow with elliptical streamlines. If $\kappa_1\kappa_2 > 0$ we have a strong flow, with hyperbolic streamlines. Simple shear, $\kappa_1\kappa_2 = 0$, is the marginal case, which we classify as weak since fluid particles separate more slowly than exponentially, and long-chain molecules would not be stretched according to our model above.

The canonical form (4.17) corresponds to

$$\underline{e} = \begin{pmatrix} 0 & \kappa \\ \kappa & 0 \end{pmatrix} \tag{4.18a}$$

$$\kappa = \tfrac{1}{2}(\kappa_1 + \kappa_2) = \surd(\dot{\gamma}^2 + \hat{\kappa}^2) \tag{4.18b}$$

and

$$\underline{\omega} = \begin{pmatrix} 0 & -\omega \\ \omega & 0 \end{pmatrix} \tag{4.19a}$$

$$\omega = \tfrac{1}{2}(\kappa_1 - \kappa_2) = \tfrac{1}{2}(\hat{\kappa}_1 - \hat{\kappa}_2). \tag{4.19b}$$

The eigenvalues of $\underline{e}$ are $\pm\ \kappa$, and we have the relation

$$\alpha^2 = \kappa^2 - \omega^2 \tag{4.20}$$

between the principal strain rate κ, the vorticity ω and the eigenvalue

$$\alpha = \sqrt{(\kappa_1 \kappa_2)} \quad (4.21)$$

of $\underline{L}$. Mackley [M7] calls this latter quantity the persistent strain rate, and gives its physical interpretation as the rate of extension (per unit length) of a non-rotating material line.

Astarita [A15,M10] discusses the use of $1/\alpha$ as a characteristic time for a flow, pointing out that it is invariant and that (4.20) is equivalent to

$$\alpha = \sqrt{(II_e - II_\omega)} \ . \quad (4.21a)$$

(We have taken a half of his value, and our definition of the invariant II is a quarter of his.) Then the dimensionless group $\lambda\alpha$ is clearly important, and the conditions for a flow to be weak, (4.14), ensure that $|\lambda\alpha| < \frac{1}{2}$. Mackley [M7] uses $\tau = \frac{1}{2}\lambda$ as a characteristic time (see the discussion in the previous section), and gives conditions for large chain extension $\alpha\tau \geq 1$ and $\alpha t >> 1$. The latter condition refers to the time spent in the flow and was mentioned in the previous section - as well as having a strong flow there must be enough time for the stretching to take place.

De Gennes [G4] considers the change in the flow (4.17) with κ_1 fixed and large as κ_2 varies, and makes estimates for the molecular transition from the coiled state for $\kappa_2 < 0$ to the extended state for $\kappa_2 > 0$. He shows that for a molecule consisting of N monomer units a change in κ_2 of order κ_1/N is enough to change the extension ratio of the molecule by a quantity of order one (i.e. typically to double the length of the molecule). Since N is large this means we have a sharp transition from the coiled state to the stretched state. However the transition is even sharper (a jump from one state to the other with a possible hysteresis effect) if we consider, say, pure stretching (with a molecule of finite extensibility) with increasing rate of strain. De Gennes calls those respectively a second order and a first order transition.

In discussing the same problem Giesekus [G7,G9] uses the quantity $\rho = \omega/\kappa$, so $\rho > 1$ gives weak (elliptical) flows and $\rho < 1$ gives strong (hyperbolic) flows. He then considers the motions of suspended rigid particles in these flows and shows how ρ and particle shape influence the stresses caused by these particles. The particle considered (for ease of computation of hydrodynamic drag forces) is a "cross dumbbell" - two rigid dumbbells of lengths ℓ_1 and ℓ_2 fixed at right angles. The shape is characterised by the parameter b = $(\ell_1^2 - \ell_2^2)/\ell_1^2 + \ell_2^2)$ and it is found that if $\rho < b$ the particles take up a fixed orientation (aligned by the flow) while for $\rho > b$ they rotate (at a non-uniform rate). The critical case $\rho = b$ corresponds to a minimum in the shear stress and zero normal stress contributions. The difference between the behaviour of flexible and rigid molecules (or models of molecules) is one that should be remembered when considering constitutive equations and critical kinematic states, though there is obviously the possibility of drawing an analogy between the coiled - stretched transition in the former case and an unoriented - oriented transition in the latter case.

4.2.2 Inhomogeneous plane flows

The converging flow at the entrance to a capillary clearly has the appearance of an extensional flow, and measurements on such flows are used to infer fluid properties in extension (section 7.1). However if we consider a steady symmetric sink flow we discover that this is not a homogeneous (or spatially uniform) flow. Marrucci and Murch [M11] consider plane and spherically symmetric sink flows, and here we shall discuss the former only. (Astarita and Marrucci [A17, pp. 101-115] give various kinematical results for spherical sink flow.) The radial velocity, for a particle at distance R(t) from the sink, is given by

$$\dot{R}(t) = -Q/R(t) \tag{4.22}$$

where $2\pi Q$ is the constant flow rate (per unit width normal to the plane) into the sink. In rectangular Cartesian coordinates we have, with $R^2 = X_1^2 + X_2^2$,

$$\dot{X}_1(t) = -QX_1(t)/R^2(t) \tag{4.22a}$$

$$\dot{X}_2(t) = -QX_2(t)/R^2(t) \tag{4.22b}$$

and hence

$$\underline{L} = \frac{Q}{R^4}\begin{pmatrix} X_1^2 - X_2^2 & 2X_1X_2 \\ 2X_1X_1 & -X_1^2 + X_2^2 \end{pmatrix} \tag{4.23}$$

This has the eigenvalues $\pm Q/R^3$ and (obviously) eigenvectors which are radial and transverse (i.e. perpendicular to the radius vector). We see therefore that if we apply criterion $\lambda\alpha > \frac{1}{2}$ locally, we shall only get chain stretching when

$$R < R_s \equiv (2\lambda Q)^{1/3}. \qquad (4.24)$$

We also note that a particle only spends a finite time, t_s, in the region where (4.24) holds, so the molecules are only stretched for this time; given by

$$t_s = \int_{R_s}^{0} \frac{dR}{\dot{R}} = \frac{R_s^2}{2Q} = \left(\frac{\lambda^2}{2Q}\right)^{1/3}. \qquad (4.25)$$

This is related to the result [M11] that the stress in this steady flow for a rubberlike liquid (or a suspension of Hookean dumbbells) never becomes infinite for a finite rate of strain, in contrast to behaviour in steady uniaxial extension and pure shear (where the molecules have as much time as they need to be aligned and stretched). It should be possible to extend this work and to estimate the amount by which chains are extended in sink flow. The results are relevant to any wedge-shaped converging flow, but clearly if the converging flow is between curved boundaries (as in the commonly observed "wine-glass stem" flow) the conclusions will be different, and we can get homogeneous extension with the correct shape of boundary.

Berry and Mackley [B12,M7] have considered a more complex set of plane flows (which they generate by means of a six-roll mill). These flows, which are also non-homogeneous, have the velocity components

$$U_1 = -2\gamma X_1 Y_2 - \omega X_2 + V_1 \qquad (4.26a)$$

$$U_2 = -\gamma(X_1^2 - X_2^2) + \omega X_1 + V_2 \qquad (4.26b)$$

and

$$\underline{L} = \begin{pmatrix} -2\gamma X_2 & -\omega-2\gamma X_1 \\ \omega-2\gamma X_1 & 2\gamma X_2 \end{pmatrix} \qquad (4.27)$$

The parameters determining the qualitative features of the flow are ω/γ, which influences $\underline{L}$ directly, and V_1/γ and V_2/γ which influence the flow by moving the stagnation points (where $\underline{U} = \underline{0}$). This is significant because $\underline{L}$ varies with position, so that V_1 and V_2 effectively alter $\underline{L}$ at the stagnation points. The discussion of this somewhat complicated three parameter family of flows is aided by the language and ideas of catastrophe theory, and the interested reader is referred to the papers cited above.

Alternatively we may consider the dynamical system defined by the nonlinear differential equations (4.26), using $U_1 = \dot{X}_1(t)$, $U_2 = \dot{X}_2(t)$ following a material particle at $(X_1(t), X_2(t))$. The trajectories of this system are just the streamlines of the flow. Then we may show that there are in general either two or four singular points for the system (stagnation points for the flow), and that there are two saddle points or three saddle points and a centre. Locally a saddle point corresponds to a strong flow (Giesekus' hyperbolic flow), and a centre to a weak flow (Giesekus' elliptic flow). The bifurcation which occurs at the transition between these types of flow involves a non-elementary singular point (or catastrophe critical point [B12]). For example when $V_1 = 0$, and taking $\gamma = 1$ for simplicity, the singular points are at

$$X_1 = -\omega/2, \; X_2 = \pm\sqrt{(3\omega^2/4 - V_2)} \tag{4.28a}$$

if $V_2 < 3\omega^2/4$ and

$$X_1 = \omega/2 \pm \sqrt{(\omega^2/4 + V_2)}, \; X_2 = 0 \tag{4.28b}$$

if $V_2 > -\omega^2/4$.

When $V_2 = 3\omega^2/4$ the two points (4.28a) and one of the points (4.28b) coincide to give a cusp catastrophe critical point at $(-\omega/2,0)$, where

$$\underline{L} = \begin{pmatrix} 0 & 0 \\ 2\omega & 0 \end{pmatrix} \tag{4.29a}$$

which has zero eigenvalues. Similarly when $V_2 = -\omega^2/4$ we have the two singular points of (4.28b) coinciding to give a fold catastrophe critical points at $(\omega/2,0)$, and here

$$\underline{L} = \begin{pmatrix} 0 & -2\omega \\ 2\omega & 0 \end{pmatrix} \tag{4.29b}$$

Thus the bifurcation or catastrophes correspond to stretching of material lines at an algebraic rather than exponential rate and are at the transition between rotating (elliptical) flows and stretching (hyperbolic) flows. The velocity gradient $\underline{L}$ has real eigenvalues

$$\alpha = \pm\sqrt{\{4\gamma^2(X_1^2 + X_2^2) - \omega^2\}} \qquad (4.30)$$

so that a stagnation point (X_1,X_2) is elliptical if it lies inside the circle

$$X_1^2 + X_2^2 = \omega^2/4\gamma^2 \qquad (4.31)$$

and hyperbolic if it lies outside.

From the experimental point of view, what is interesting is that the dissolution of small amounts of polymer in the liquid in the six-roll mill affects the streamline pattern qualitatively, to the extent that four stagnation points were not obtained (under circumstances when they were obtained for the Newtonian solvent). Berry and Mackley suggest that the dissolved polymer is extended in parts of the flow field and that therefore, because large stresses cannot be provided, high elongational strain rates are reduced. Observations of flow-induced birefringence lend support to the belief that the polymer chains are stretched in the expected regions of the flow.

4.3 ELONGATIONAL FLOWS

The examples we have been considering in the previous chapters may be generalised, and we consider three stages of this generalisation here. First we shall make an obvious generalisation to steady flows and then we shall consider unsteady flows. The problem of finding a frame-indifferent definition will lend us to a discussion of stretching and rotation. In the final sub-section we briefly review motions with constant stretch history - a class of motions which includes both steady simple shear and steady elongation. This helps to set our flow classifications in a slightly wider context; for a more general discussion the reader is referred to Biermann [B13].

4.3.1 Steady rectilinear extension

This flow, also called simple extensional flow, is a flow with velocity components

$$U_i = e_i X_i \; ; \quad i = 1,2,3 \qquad (4.32)$$

in some fixed Cartesian coordinate system [C31]. Clearly the rate of strain tensor $\underline{e}$ is diagonal and has principal values e_1, e_2 and e_3. Coleman and Noll [C31] prove that for a simple fluid

$$\underline{p} = g\underline{I} + h\underline{e} + \ell\underline{e}^2 \tag{4.33}$$

where g, h and ℓ are functions of the invariants of $\underline{e}$. The function g does not affect the differences between extra-stress components which we measure but is needed for consistency with the constitutive equations we use. The proof of (4.33) rests on representation theorems for isotropic functions; we discuss this below (section 4.3.5). Instead of the tensor equation (4.33) we may express the result in terms of the components of $\underline{p}$ and $\underline{e}$

$$p_{ii} = f(e_i, e_j, e_k) \tag{4.34}$$

where (i,j,k) are the integers (1,2,3) in any order and

$$f(e_i, e_k, e_j) = f(e_i, e_j, e_k). \tag{4.35}$$

Clearly $p_{ij} = 0$ for $i \neq j$, and f is related to g, h and ℓ by

$$f(e_i, e_j, e_k) = g + he_i + \ell e_i^2 \tag{4.36}$$

and the symmetry (4.35) follows from the symmetry in the definitions of the invariants of $\underline{e}$. The two forms (4.33) and (4.34) are exactly equivalent, and there are no restrictions on the functions f, g, h and ℓ other than (4.35) (and the relation (4.36)).

Clearly for the Newtonian liquid $h = 2\mu$, $g = \ell = 0$ and we may deduce these functions (and f) for other constitutive equations. For example by substituting (4.33) into (2.25) and demanding that it be satisfied identically we obtain for the Oldroyd eight-constant model

$$h = 2\eta_o\left(\frac{1+4\tau_2 II_e + 4\mu_2(3\tau_1+\mu_1^2)III_e}{1+4\tau_1 II_e + 4\mu_1(3\tau_1+\mu_1^2)III_e}\right) \tag{4.37a}$$

$$\ell = 4\eta_o\left(\frac{(\mu_1-\mu_2)-2(\mu_1\nu_2-\mu_2\nu_1)(3\mu_o-2\mu_1)II_e}{1+4\tau_1 II_e + 4\mu_1(3\tau_1+\mu_1^2)\ III_e}\right) \tag{4.37b}$$

where

$$\tau_1 = \sigma_1 - \lambda_1^2, \quad \tau_2 = \sigma_2 - \lambda_1\lambda_2 \tag{4.37c,d}$$

and σ_1 and σ_2 are defined in (2.51). The second term in the numerator of ℓ can be expressed in terms of τ_1, τ_2, μ_1 and μ_2 so h and ℓ only involve five constants.

For the particular flows discussed in section 2.2 we have the relations

$$\eta_T = \frac{3}{2} h \left(\frac{3\dot{\gamma}^2}{4}, \frac{\dot{\gamma}^3}{4} \right) + \frac{3\dot{\gamma}}{4} \ell \left(\frac{3\dot{\gamma}^2}{4}, \frac{\dot{\gamma}^3}{4} \right), \quad (4.38a)$$

$$\eta_{eb} = 3h(3\dot{\Gamma}^2, -2\dot{\Gamma}^3) - 3\dot{\Gamma}\ \ell(3\dot{\Gamma}^2, -2\dot{\Gamma}^3), \quad (4.38b)$$

$$\eta_{sb} = 2h(\dot{\varepsilon}^2, 0)\ , \quad (4.38c)$$

$$\eta^x_{sb} = h(\dot{\varepsilon}^2, 0) - \dot{\varepsilon}\ \ell(\dot{\varepsilon}^2, 0)\ . \quad (4.38d)$$

As well as providing a check on the consistency of (4.37), (2.52) and (2.53), these relations show how limited is the information obtainable from particular experiments in dealing with functions of the invariants (cf. Lodge [L17], p.224]). Related calculations have been carried out for the upper convected Maxwell model by Stevenson [S19] who also proposes a classification of elongational flows based on the fact that $e_1 + e_2 + e_3 = 0$ for an incompressible fluid. In essence the two independent quantities give a magnitude of the rate of elongation and a ratio which characterises the type of flow (e.g. uniaxial, strip biaxial, equal biaxial).

Equation (4.33) may be identified as the Reiner-Rivlin equation and is the elongational analogue of the Criminale-Ericksen-Filbey equation in steady shearing flow [B18]. The functions h and ℓ are the analogue of the three viscometric functions (cf. section 2.2 above and [D3,C30,B18,A17]). For a general simple fluid there is no general relation between h and ℓ and the viscometric functions, and this observation has an important practical counterpart in the observation [A5] that the large relaxation time part of the spectrum plays a significant role in elongational flow, but not in shear. An associated point is made by Roscoe [R12] who calculates the material functions for a third-order (differential) constitutive equation in steady simple shear, small amplitude oscillatory shear, and steady elongation. He shows how the combination of material constants appearing in the expression for elongational viscosity, η_T, may be obtained from the shear flow measurements. The second-order results (cf. section 2.2)

$$\eta(\kappa) = \eta \tag{4.39a}$$

$$\Psi_1(\kappa) + 2\Psi_2(\kappa) = 2(\chi-\theta) \tag{4.39b}$$

allow

$$\eta_T(\dot{\gamma}) = 3\eta + 3(\chi-\theta)\dot{\gamma} \ . \tag{4.39c}$$

to be calculated from viscometric functions alone. For the third-order fluid this is not possible since the coefficient of $\underline{A}_3$ cannot be determined from viscometric measurements (where $\underline{A}_3 \equiv 0$), and the oscillatory shear measurements are needed. These would not suffice for higher-order fluids, since in neither steady nor small amplitude oscillatory shear does the term $(\underline{A}_1 \cdot \underline{A}_3 + \underline{A}_3 \cdot \underline{A}_1)$ affect the stresses, although it does in elongation.

We may make a contrary observation for the Oldroyd model, where a full knowledge of the viscometric functions (2.51) allows the constants η_o, λ_1, $\lambda_2, \mu_1, \mu_2, \sigma_1$ and σ_2 to be determined. These suffice to calculate h and ℓ from (4.37) and hence the stresses in any steady extension. We must emphasise that this result is a consequence of the choice of model. Indeed connections between shearing and elongational flow behaviour can only be made if a specific model is chosen. An attempt to deduce an approximate relation for a general simple fluid has been made by Zahorski [Z1]. However it is doubtful that Zahorski's result has any wider validity than the approximation of slow and smooth flow which leads to the second-order fluid (2.40). For the upper convected Maxwell model the error in Zahorski's result is in fact $O(\dot{\gamma}^2)$.

The question of whether a steady elongational flow can be realised depends on several factors, some of which we have discussed above, notably (in section 2.3) the stress growth when the flow is started. The stability of such a flow also merits discussion and here we consider a third factor, namely whether such a flow is dynamically possible with realistic boundary conditions and body forces. Coleman and Noll [C31] show that the homogeneous extension of a circular cylinder in the absence of body forces depends on having a position-dependent radial stress acting on the free surface of the cylinder unless inertia is negligible. They show that, if the end of the cylinder at $X_1 = 0$ is stationary and the other end is at $X_1 = L(t)$, then the axial stresses applied are

$$T(0) = 3\dot{\gamma}h/2 + 3\dot{\gamma}^2\ell/4 + \tfrac{1}{2}\rho\dot{\gamma}^2(R^2/4 - L^2/3) \quad (4.40a)$$

$$T(L) = 3\dot{\gamma}h/2 + 3\dot{\gamma}^2\ell/4 + \tfrac{1}{2}\rho\dot{\gamma}^2(R^2/4 + 2L^2/3) \quad (4.40b)$$

where $R(t)$ is the radius of the cylinder, $T(X)$ is the axial stress and h and ℓ are the functions defined in (4.33). The radial stress at the free surface is

$$P(X_1) = \tfrac{1}{2}\rho\dot{\gamma}^2(X_1^2 - L^2/3). \quad (4.40c)$$

If this is not provided then we may expect the cylinder to bulge for $0 \leq X_1 < L/\sqrt{3}$, and to narrow for $L/\sqrt{3} < X_1 \leq L$. Balmer [B5] discusses the influence of this in the tubeless siphon flow and claims to observe departures of the shape of the column from the ideal which are attributable to this effect. Since there are many other non-ideal features of this flow, this conclusion is not incontrovertible. (See also Stevenson [S18] for a discussion of the effects of surface tension and gravity.)

4.3.2 Steady extension

A generalisation of the definition of steady rectilinear extension has been proposed by Coleman [C34], in which the requirement that the fixed coordinate system by Cartesian is relaxed, and the rate of strain tensor is diagonal when referred to some fixed orthogonal coordinate system. The sink flow discussed in section 4.2.2 is probably the simplest example of a non-rectilinear steady extension. We notice two rather simple and obvious things about this example; if the flow is steady it cannot be homogeneous (i.e. spatially uniform) so that it is not always possible for an incompressible fluid to undergo a prescribed homogeneous steady extension; the flow has straight streamlines, whereas in rectilinear steady extension the streamlines are curved, so that the nomenclature does have a tendency to lead to confusion if care is not taken. Furthermore equation (4.32) cannot be used to define steady extension; if we consider cylindrical polar coordinates (r,θ,z) and use (4.32) we have velocity components $(c_1 r, c_2\theta, c_3 z)$ and the velocity gradient tensor

$$\underline{L} = \begin{pmatrix} c_1 & -c_2\theta/r & 0 \\ 0 & c_1+c_2/r & 0 \\ 0 & 0 & c_3 \end{pmatrix} \quad (4.41)$$

so that $\underline{e}$ $(= \frac{1}{2}(\underline{L} + \underline{L}^T))$ is not diagonal. In fact Coleman's definition is expressed in terms of a stretch tensor rather than a rate of strain tensor, and we discuss this below (section 4.3.4).

Coleman's discussion of extensional flows [C34] is not restricted to steady flows, so we defer discussion of his results to section 4.3.3 below. The particular examples he gives are the inflation and stretching of a circular tube (using cylindrical polar coordinates), inflation of a spherical shell (using spherical polar coordinates) and bending of a block into a cylindrical wedge (using Cartesian and cylindrical polar coordinates). Balmer and Astarita [B4] discuss the first two of these in more detail, seeking conditions under which the flow may be a motion with constant stretch history (section 4.3.5 below). For inflation or deflation of a tube a material point undergoing purely radial motion with exponential time dependence has its position given by

$$r^2 = \rho^2 + \rho_*^2 (\exp(2at) - 1) \tag{4.42}$$

where the particle at radius r at time t was at radius ρ at time zero, and a and ρ_* are constants. Then the radial velocity is

$$u = a\rho_*^2 \exp(2at)/r \tag{4.43}$$

and the principal values of the rate of strain tensor $\underline{e}$ are

$$e_1 = (\partial u/\partial r)_t = -a\rho_*^2 \exp(2at)/r^2 \tag{4.44a}$$

$$e_2 = u/r = a\rho_*^2 \exp(2at)/r^2 \tag{4.44b}$$

and

$$e_3 = 0 \tag{4.44c}$$

If $\rho = \rho_*$ we can eliminate t between (4.42) and (4.43) to obtain u = ar. Note, however, that this does not imply that $e_1 = a$; (4.44a) correctly gives $e_1 = -a$ in this case. It is wrong to differentiate u = ar with respect to r to get a strain rate because u = ar only on a cylinder of radius r which is itself a function of time, $r = \rho_* \exp(at)$.

Thus it is possible to have a constant strain rate on this one material surface, but the requirements of continuity (and incompressibility) do not allow a purely radial motion with a constant strain rate (or even a motion with constant stretch history) throughout a thick-walled cylinder. Contrast with this the flow discussed in section 4.2.2, where the constant flow rate of the steady flow leads to a different dependence on the radial coordinate r. An analogous result is obtained for the inflation of a sphere. A combination of axial stretching and radial contraction of a tube is just the uniform axisymmetric stretching we have discussed in earlier chapters and is, not surprisingly, expressible in cylindrical polar coordinates as a flow with a constant strain rate.

Balmer and Astarita [B4] go on to discuss the dynamics of radial expansion of a thin-walled tube, axial stretching and radial contraction of a tube and inflation or deflation of a thin-walled spherical balloon. In the absence of inertia they find that these motions require a uniform pressure (independent of position) inside the tube or balloon and so are realisable in principle, as slow flows at least. Chung and Stevenson [C16] develop this idea further and discuss the influence of gravity and inertia, as well as working out detailed predictions for the upper convected Maxwell model. (The experiment of stretching and inflating a cylindrical tube, and at the same time measuring both the kinematics and the forces acting has not been carried out.) Experiments on inflation alone have been discussed in section 3.2 above, for thin sheets of highly viscous polymer.

The spherical geometry has also been used by Pearson and Middleman [P1] in a study of the collapse of a gas bubble in an elasticoviscous liquid. Their calculations of the kinematics and dynamics do not use an assumption of exponential time dependence but find that this is observed in their experiments over much of the time of observation. Surface tension is not ignored, but otherwise the treatment is equivalent to that of Balmer and Astarita. The necessity of integrating the stress over the whole flow field (from $r = R(t)$ at the bubble surface out to $r = \infty$ to obtain an expression for the pressure in the bubble) means that, as with spinning and other nonuniform flows, it is difficult to get any useful quantitative information without choosing some particular constitutive equation. A comparison is made [P1] between elongational viscosity functions and an apparent elongational viscosity based on the strain rate at the bubble surface. This is more reasonable than it would be for the sink flow (with constant flow rate) discussed earlier since the observed constant strain rate at that material surface corresponds to a decreasing volume flow rate out of the bubble.

Of course the result of Balmer and Astarita [B4] tells us that the strain rate at any other material surface is not constant, so the integration referred to above implies that the calculated pressure inside the bubble is influenced by flow of material with a decreasing strain rate. We may show this explicitly, taking

$$e_2 = e_3 = \dot{\Gamma} \equiv - \dot{R}(t)/R(t) , \tag{4.45}$$

and $e_1 = -2\dot{\Gamma}$ as principal rates of strain at the bubble surface (transversely and radially). If $\dot{\Gamma}$ is constant then clearly

$$R(t) = R_o \exp(-\dot{\Gamma}t) . \tag{4.46}$$

At any other radius, r, the transverse strain rate is (by continuity)

$$e_2 = e_3 = \dot{\Gamma}R(t)^3/r^3 . \tag{4.47}$$

The material surface which is at $\rho(t)$, and was initially at ρ_o, is determined also by conservation of mass and (analogously to (4.42) for the cylindrical geometry)

$$\rho(t)^3 - R(t)^3 = \rho_o^3 - R_o^3 \tag{4.48}$$

Hence the transverse strain rate at this material surface is, using (4.46) and (4.48) in (4.47)

$$e_2 = e_3 = \dot{\Gamma}/[1 + \exp(3\dot{\Gamma}t)\{(\rho_o^3/R_o^3) - 1\}]. \tag{4.49}$$

which decreases with time since $\dot{\Gamma} > 0$ a collapsing bubble and $\rho_o > R_o$.

4.3.3 Unsteady elongation

Coleman [C34] generalises the above to deal with time-varying (unsteady) elongational flows where the principal rates of strain e_1, e_2 and e_3 are not constant but are functions of time. He considers pure stretching flows, where the principal directions of the right stretch tensor (see below, section 4.3.4) are constant in time, and deduces from considerations of symmetry the special form which the constitutive equation of the simple material takes. The special case of incompressible materials is considered by Coleman and Dill [C35] and they also discuss the stability of the motions of inflation of a circular tube and of a spherical shell. The discussion of the stability of solutions appears to be more appropriate to viscoelastic solids than to elasticoviscous fluids and is not pursued here.

Similar results have been obtained by Lodge [L17] who uses the term shear-free flow, defined to be a flow in which there is a body coordinate system which is orthogonal for all time and by Phillips [P23] for the start-up of steady elongation. Slattery [S9], Zahorski [Z2] and Chow and Carroll [C15] consider respectively "unsteady relative extension", "motions with proportional stretch history" and "motions of proportional extension" and obtain results more like (4.33) for their more restricted kinematics. The general result of Coleman and of Lodge is that a single (scalar-valued) functional, p, suffices to give the principal stresses, p_{ii};

$$p_{11}(t) = \underset{t'=-\infty}{\overset{t}{p}} \{e_1(t'),\ e_2(t'),\ e_3(t')\}, \tag{4.50a}$$

$$p_{22}(t) = \underset{t'=-\infty}{\overset{t}{p}} \{e_2(t'),\ e_3(t'),\ e_1(t')\}, \tag{4.50b}$$

$$p_{33}(t) = \underset{t'=-\infty}{\overset{t}{p}} \{e_3(t'),\ e_1(t'),\ e_2(t')\} \tag{4.50c}$$

and the functional p is symmetric in its last two arguments

$$p\{e_1,\ e_2,\ e_3\} = p\{e_1,\ e_3,\ e_2\}. \tag{4.50d}$$

For an incompressible material this can be simplified by using the fact that $e_1 + e_2 + e_3 \equiv 0$, and we obtain [C35] the extra-stress differences

$$p_{11}(t) - p_{22}(t) = \underset{t'=-\infty}{\overset{t}{f}} \{e_1(t'),\ e_2(t')\} \tag{4.51a}$$

$$p_{11}(t) - p_{33}(t) = \underset{t'=-\infty}{\overset{t}{f}} \{e_1(t'),\ e_3(t')\} \tag{4.51b}$$

where the functional f satisfies

$$f\{e_1,\ e_2\} = -f\{e_2,\ e_1\} = f\{e_1,\ e_3\} - f\{e_2,\ e_3\}. \tag{4.51c,d}$$

As an example, using Cartesian coordinates, for the rubberlike liquid we first obtain from

$$U_i(t'') = e_i(t'')\, X_i(t'') \tag{4.52}$$

(for a material particle at position $\underline{X}(t'')$ at time t") and the identity

$$U_i(t'') = d\, X_i(t'')/dt'' \tag{4.53}$$

the principal stretch ratios

$$\alpha_i(t,t') \equiv X_i(t)/X_i(t') = \exp \int_{t'}^{t} e_i(t'')dt'' \tag{4.54}$$

and hence have

$$\left(C_t^{-1}(t')\right)_{ii} = \alpha_i(t,t')^2 \; ; \; i = 1,2,3 \; . \tag{4.55}$$

Then the functional f is given by

$$f\{e_1, e_2\} = \int_{-\infty}^{t} m(t-t')\left(\alpha_1(t,t')^2 - \alpha_2(t,t')^2\right) dt' \tag{4.56}$$

together with (4.54). For the upper convected Maxwell model we replace m(t-t') by the particular memory function $(G/\lambda)\exp\{-(t-t')/\lambda\}$, as usual.

Coleman and Dill extend these results, which refer to isotropic materials, to transversely isotropic materials with axis of symmetry in the X_1 direction, and state that all of (4.51) except (4.51d) will still hold. In addition they give a generalised dissipation inequality for their stress functional and a similar free energy functional. A further result concerns the principal directions of the stress in more general extensional flows where these may rotate, unlike the example we have been considering (4.52) where the fixed Cartesian coordinate axes are the principal directions of stress, strain and rate of strain. Lodge [L17] states this is the form that the principal directions of stress are everywhere orthogonal to the surfaces of the shear-free flow body coordinate system.

Astarita and Marrucci [A17] give the equivalent result, that the principal directions of stress coincide with the principal directions of stretch. This follows the work of Coleman [C34], with the important distinction that Coleman deals with solids (in the sense that he measures deformation and stretch relative to a fixed reference configuration) rather than liquids (Astarita and Marrucci use the present configuration as the reference configuration). We defer discussion of these results, for which we need the polar decomposition theorem, to section 4.3.4 below.

The more specialised results of Slattery, Zahorski, Chow and Carroll, and Phillips all depend on the consideration of a more restricted class of flows in which the principal rates of strain e_1, e_2, e_3 all have the same time dependence, i.e. (4.32) is replaced by

$$U_i = e_i k(t) X_i ; \; i = 1,2,3 \tag{4.57}$$

where k(t) is an arbitrary function of time which Slattery and Phillips restrict to be, respectively, an exponential and a step function. The basic result in all cases is a relation

$$\underline{p}(t) = H\underline{e}(t) + L\underline{e}^2(t) \; . \tag{4.58}$$

Here H and L are functions of invariants of $\underline{e}$ (like h and ℓ in (4.33)) and of time (through the function k(t)). Balmer [B5] claims that all unsteady extensional motions are of the form (4.37) so that they are d'Alembert motions, i.e. motions in which the velocity field is expressible as the product of a function of position and a function of time (or is separable). This does not appear to be the case if we start from Coleman's definition and would exclude the motion

$$U_1 = a(t) X_1 \tag{4.59a}$$

$$U_2 = (b(t) - a(t)) X_2 \tag{4.59b}$$

$$U_3 = -b(t) X_3$$

which is in fact of the form cited by Balmer in a footnote as being equivalent to (4.57), though if a(t) and b(t) are different non-zero functions of time the equivalence is not at all obvious.

4.3.4 Stretching and rotation

The basic kinematics of deformation are discussed in many textbooks on continuum mechanics, e.g. [A17,T25] and we shall quote results here without proof. In dealing with fluids we have been using tensors expressing deformation relative to the configuration of the material at the present time t, as in (2.18,2.19), while in much basic continuum mechanics a fixed reference configuration (Ξ_1,Ξ_2,Ξ_3) is used, as in Coleman's definition of an elongational flow [C34]. We define the deformation gradient $\underline{F}(t)$ by its Cartesian components

$$\{\underline{F}(t)\}_{ij} = \partial X_i/\partial \Xi_j \tag{4.60}$$

and the relative deformation gradient $\underline{F}_t(t')$ by

$$\{\underline{F}_t(t')\}_{ij} = \partial X'_i/\partial X_j \ . \tag{4.61}$$

These are related by

$$\underline{F}(t') = \underline{F}_t(t') \cdot \underline{F}(t) \tag{4.62}$$

(the chain rule of partial differentiation).

The basic result we need is the polar decomposition theorem which states that, since $\underline{F}(t)$ is invertible (its determinant is one for an incompressible material), it possesses a unique decomposition

$$\underline{F}(t) = \underline{R}(t) \cdot \underline{U}(t) \tag{4.63}$$

into the product of an orthogonal tensor $\underline{R}(t)$ (which represents a rotation) and a positive definite symmetric tensor $\underline{U}(t)$ (the right stretch tensor). Similarly

$$\underline{F}_t(t') = \underline{R}_t(t') \cdot \underline{U}_t(t') \tag{4.64}$$

determines uniquely the orthogonal tensor $\underline{R}_t(t')$ and the positive definite symmetric tensor $\underline{U}_t(t')$. Coleman [C34] defines an extensional flow as one for which there exists a fixed orthogonal coordinate system with respect to which $\underline{U}(t')$ has a diagonal matrix of components. Our purpose now is to relate this to the approaches involving relative deformation or rate of strain. We use the results [T25]

$$\underline{e}(t) = \dot{\underline{U}}_t(t) = \tfrac{1}{2}\underline{R}(t)\cdot\{\dot{\underline{U}}(t)\cdot\underline{U}^{-1}(t) + \underline{U}^{-1}(t)\cdot\dot{\underline{U}}(t)\}\cdot\underline{R}^T(t) \tag{4.65a,b}$$

$$\underline{\omega}(t) = -\dot{\underline{R}}_t(t) = -\dot{\underline{R}}(t)\cdot\underline{R}^T(t) - \tfrac{1}{2}\underline{R}(t)\{\dot{\underline{U}}(t)\cdot\underline{U}^{-1}(t) - \underline{U}^{-1}(t)\cdot\dot{\underline{U}}(t)\}\cdot\underline{R}^T(t) \tag{4.66a,b}$$

where $\dot{\underline{F}}_t(t)$ denotes $\partial\underline{F}_t(t')/\partial t'$, evaluated at $t' = t$. Note that we have defined the vorticity tensor using the sign convention common in hydrodynamics (2.14) rather than continuum mechanics [A17,T25] so that the relationships obtained there have to be adjusted (their $\underline{W}$ is $-\underline{\omega}$ in our notation).

There are no general explicit relations between $\underline{R}(t)$, $\underline{R}(t')$ and $\underline{R}_t(t')$ or $\underline{U}(t)$, $\underline{U}(t')$ and $\underline{U}_t(t')$, similar to (4.62) for $\underline{F}$. However if a motion is an extension $\underline{U}(t)$ and $\underline{U}(t')$ commute (this is obvious if we use the time-independent coordinate system in which $\underline{U}(t)$ is diagonal, and hence may be shown for any coordinate system). Then, because in addition $\underline{U}$ is symmetric, the product $\underline{U}(t')\cdot\underline{U}^{-1}(t)$ is symmetric and if we define the matrices

$$\hat{\underline{U}}(t,t') = \underline{R}(t)\cdot\underline{U}(t')\cdot\underline{U}^{-1}(t)\cdot\underline{R}^T(t)\,, \tag{4.67a}$$

$$\hat{\underline{R}}(t,t') = \underline{R}(t')\cdot\underline{R}^T(t) \tag{4.67b}$$

these are respectively positive definite symmetric and orthogonal. It is an easy matter to verify that

$$\hat{\underline{R}}(t,t')\cdot\hat{\underline{U}}(t,t') = \underline{F}(t')\cdot\underline{F}^{-1}(t) = \underline{F}_t(t') \tag{4.68}$$

so that (4.67) does give the polar decomposition of $\underline{F}_t(t')$ and $\underline{U}_t(t') = \hat{\underline{U}}(t,t')$, $\underline{R}_t(t') = \hat{\underline{R}}(t,t')$. It must be emphasised that this deduction depends on the proof that $\hat{\underline{U}}(t,t')$ is symmetric. Without this the definition (4.67) can still be made and (4.68) is still true, but if $\hat{\underline{U}}(t,t')$ is not symmetric then (4.68) is not the polar decomposition of $\underline{F}_t(t')$. The matrix $\underline{U}(t)$ for simple shear is readily seen to be one for which $\hat{\underline{U}}(t,t')$ calculated using (4.67a) is not symmetric.

We note that if $\underline{U}(t)$ is made diagonal, by the appropriate choice of coordinates, $\underline{U}_t(t')$, given by (4.67a), is not. However if we change frame using

$$\underline{X}^* = \underline{Q}(t')\cdot\underline{X} \tag{4.69}$$

then [T11]

$$\underline{R}_t^*(t') = \underline{Q}(t')\cdot\underline{R}_t(t')\cdot\underline{Q}^T(t)\,, \tag{4.70a}$$

$$\underline{U}_t^*(t') = \underline{Q}(t)\cdot\underline{U}_t(t')\cdot\underline{Q}^T(t) \tag{4.70b}$$

and taking $\underline{Q}(t) = \underline{R}^T(t)$ gives the diagonal tensor

$$\underline{U}_t^*(t') = \underline{U}(t')\cdot\underline{U}^{-1}(t)\,. \tag{4.71a}$$

In addition, from (4.65a),

$$\underline{e}^*(t) = \dot{\underline{U}}(t)\cdot\underline{U}^{-1}(t) \tag{4.71b}$$

which is also diagonal and

$$\underline{\omega}^{*}(t) = \underline{O} \tag{4.71c}$$

from (4.66a) and, using (4.67b) and (4.70a),

$$\underline{R}_{t}^{*}(t') = \underline{I} \, . \tag{4.71d}$$

We see that the principal directions of $\underline{U}_t(t')$ and $\underline{R}(t)\cdot\underline{u}_i$ (if $\underline{u}_i$ are the principal directions of $\underline{U}(t)$) so the result of Coleman [C34], that the principal direction of the stress tensor is an extensional flow are $\underline{R}(t)\cdot\underline{u}_i$, is exactly equivalent to that of Astarita and Marrucci [A17] that the stress has the same principal directions as the relative right stretch tensor, $\underline{U}_t(t')$. We see also that for an extensional flow as defined by Coleman it is always possible to choose a frame (and a set of axes) in which $\underline{U}_t(t')$ and $\underline{e}(t)$ are both diagonal, so that the study of (4.32) and (4.52) involves no loss of generality.

Note that Coleman's definition of an extensional flow is frame-indifferent, whether expressed in terms of the principal axes of $\underline{U}(t)$ or of $\underline{U}_t(t')$. If $\underline{U}_t(t')$ is diagonal (in the unstarred frame) in some coordinate system then the change of frame represented by $\underline{Q}(t')$ has the same effect (4.70b) on $\underline{U}_t(t')$ as a rotation of axes given by $\underline{Q}(t)$ and hence principal axes of $\underline{U}_t^{*}(t')$ are rotated by a fixed amount (independent of t') from those of $\underline{U}_t(t')$. This is more immediately evident in the body tensor formalism of Lodge [L17].

We next study the relation between elongational flows and two classes of flow whose definitions are not frame-indifferent. An <u>irrotational flow</u> is one in which the vorticity tensor, $\underline{\omega}$, is identically zero. A <u>pure stretching flow</u> is one in which the rotation tensor, $\underline{R}(t)$, is the identity tensor, $\underline{I}$, for all time, t. It is clear from (4.66b) that a pure stretching flow is irrotational if and only if the tensors $\dot{\underline{U}}(t)$ and $\underline{U}^{-1}(t)$ commute. Since this is true for an extensional flow we have

<u>Result (a)</u> An extensional flow which is a pure stretching flow is necessarily irrotational

<u>Result (b)</u> An irrotational extensional flow is necessarily a pure stretching flow.

There is not a complete formal proof of

<u>Conjecture (c)</u> An irrotational pure stretching flow is an extensional flow.

In addition it is relatively easy to show that

<u>Result (d)</u> Under the change of frame (4.69) with $\underline{Q}(t) = \underline{R}^{T}(t)$ an extensional

flow becomes both a pure stretching flow,

$$\underline{R}^*(t) = \underline{Q}(t)\cdot\underline{R}(t) = \underline{R}^T(t)\cdot\underline{R}(t) = \underline{I} \tag{4.72}$$

and an irrotational flow

$$\underline{\omega}^*(t) = \underline{Q}(t)\cdot\underline{\omega}(t)\cdot\underline{Q}^T(t) - \dot{\underline{Q}}(t)\cdot\underline{Q}^T(t)$$

$$= -\underline{R}^T(t)\cdot\dot{\underline{R}}(t)\cdot\underline{R}^T(t)\cdot\underline{R}(t) - \dot{\underline{R}}^T(t)\cdot\underline{R}(t) = \underline{0} . \tag{4.73}$$

(The transformations of $\underline{\omega}$ and $\underline{R}$ under a change of frame are deduced in [T25].) Hence we make

Conjecture (e) A flow is an extensional flow if and only if there is a frame of reference in which it is both irrotational and a pure stretching flow.

The proof of (e) requires (c) and (d), so we conclude this section with a proof of (c) under the restriction that $\underline{U}(t)$ has distinct eigenvalues.

From (4.66b) if $\underline{\omega}(t) = \underline{0}$ and $\underline{R}(t) = \underline{I}$ (the hypotheses of (c)) then

$$\dot{\underline{U}}(t)\cdot\underline{U}^{-1}(t) = \underline{U}^{-1}(t)\cdot\dot{\underline{U}}(t) . \tag{4.74}$$

For positive definite symmetric $\underline{U}(t)$ with distinct eigenvalues we have

$$\underline{U}(t) = \underline{N}(t)\cdot\underline{\Lambda}(t)\ \underline{N}^T(t) \tag{4.75}$$

with diagonal $\underline{\Lambda}(t)$ whose elements are the real positive eigenvalues of $\underline{U}(t)$ (the right principal stretches) and real orthogonal $\underline{N}(t)$. Then $\underline{N}^T(t)\cdot\dot{\underline{N}}(t)$ is antisymmetric and we may use this fact to show that $\dot{\underline{N}}(t)$ must be zero if $\dot{\underline{U}}(t)$ and $\underline{U}^{-1}(t)$ commute. Putting (4.75) in (4.74) gives

$$\dot{\underline{N}}(t)\cdot\underline{N}^T(t) + \underline{N}(t)\{\dot{\underline{\Lambda}}(t)\cdot\underline{\Lambda}^{-1}(t) + \underline{\Lambda}(t)\cdot\dot{\underline{N}}^T(t)\cdot\underline{N}(t)\cdot\underline{\Lambda}^{-1}(t)\}\cdot\underline{N}^T(t)$$

$$= \underline{N}(t)\cdot\dot{\underline{N}}^T(t) + \underline{N}(t)\cdot\{\underline{\Lambda}^{-1}(t)\cdot\dot{\underline{\Lambda}}(t) + \underline{\Lambda}^{-1}(t)\cdot\underline{N}^T(t)\cdot\dot{\underline{N}}(t)\underline{\Lambda}(t)\}\cdot\underline{N}^T(t) \tag{4.76a}$$

Clearly $\dot{\underline{\Lambda}}(t)$ and $\underline{\Lambda}^{-1}(t)$ commute, so we have, using $\underline{N}(t)\cdot\underline{N}^T(t) = \underline{I}$ and $\dot{\underline{N}}(t)\cdot\underline{N}^T(t) = -\underline{N}(t)\cdot\dot{\underline{N}}^T(t)$,

$$\underline{N}(t)\{2\underline{N}^T(t)\cdot\dot{\underline{N}}(t) - \underline{\Lambda}(t)\cdot\underline{N}^T(t)\cdot\dot{\underline{N}}(t)\cdot\underline{\Lambda}^{-1}(t) - \underline{\Lambda}^{-1}(t)\cdot\underline{N}^T(t)\dot{\underline{N}}(t)\cdot\underline{\Lambda}(t)\}\underline{N}^T(t) = \underline{0} \tag{4.76b}$$

and hence the term inside the brackets { } must be zero. Putting

$$\underline{\Lambda}(t) = \begin{bmatrix} \lambda_1 & 0 & 0 \\ 0 & \lambda_2 & 0 \\ 0 & 0 & \lambda_3 \end{bmatrix} \quad \text{and} \quad \underline{N}^T(t)\cdot\dot{\underline{N}}(t) = \begin{bmatrix} 0 & a & b \\ -a & 0 & c \\ -b & -c & 0 \end{bmatrix} \tag{4.77}$$

gives three equations

$$a\left(2 - \frac{\lambda_1}{\lambda_2} - \frac{\lambda_2}{\lambda_1}\right) = b\left(2 - \frac{\lambda_1}{\lambda_3} - \frac{\lambda_3}{\lambda_1}\right) = c\left(2 - \frac{\lambda_2}{\lambda_3} - \frac{\lambda_3}{\lambda_2}\right) = 0 \tag{4.78}$$

and hence if $\lambda_1 \neq \lambda_2$, $\lambda_2 \neq \lambda_3$ and $\lambda_3 \neq \lambda_1$ then $a = b = c = 0$ and so $\dot{\underline{N}}(t) = \underline{0}$.

Hence there is a fixed set of axes in which $\underline{U}(t)$ takes the diagonal form $\underline{\Lambda}(t)$ and the motion is an extensional flow. If $\underline{U}(t)$ has two equal eigenvalues $\underline{N}(t)$ is not uniquely determined and it is very likely that a constant $\underline{N}(t)$ can still be found. The case of three equal eigenvalues corresponds to no stretching, since $\det(\underline{U}(t)) = 1$. Hence conjectures (c) and (e) are believed by the author to be true.

As illustrations of these ideas and the associated algebra, the reader is invited to consider two motions for which we use $t = 0$ as the reference state. Example 1, following Chaure [C12], is given by Cartesian velocity components

$$U_1 = \tfrac{1}{2}\kappa(X_1 \sin\kappa t + X_2 \cos\kappa t) \ , \tag{4.79a}$$

$$U_2 = \tfrac{1}{2}\kappa(X_1 \cos\kappa t + X_2 \sin\kappa t) \ , \tag{4.79b}$$

$$U_3 = 0 \tag{4.79c}$$

whence $\underline{\omega}(t) = 0$, and the change of frame which gives $\underline{R}_o^*(t) = \underline{I}$ leads to

$$\underline{\omega}^*(t) = \tfrac{1}{2}\kappa(2 + \tfrac{1}{4}\kappa^2 t^2) \begin{bmatrix} 0 & 1 & 0 \\ -1 & 0 & 0 \\ 0 & 0 & 0 \end{bmatrix} \ . \tag{4.80}$$

Example 2 (see the discussion at the end of section 4.1.2) has

$$U_1 = \alpha X_1 - \beta X_2 \tag{4.81a}$$

$$U_2 = \beta X_1 + \alpha X_2 \tag{4.81b}$$

$$U_3 = -2\alpha X_3 \ . \tag{4.81c}$$

Here $\underline{\omega}(t) \neq 0$ and $\underline{R}_o(t) \neq \underline{I}$, and the one change of frame alters both inequalities to equalities. The right stretch tensor $\underline{U}_o(t)$ is diagonal, which was not the case for example 1.

4.3.5 Motions with constant stretch history

In concluding this chapter we seek to place elongational flows in a wider classification, and to discuss the attempt which has been made to define a class of flows which are "steady" in a rheologically significant sense of the word. We interpret this as requiring that the function describing the deformation history of a material particle should not depend on present time, t, although the values of the function at past times, t', will generally depend on the time lapse t-t'. In its simplest form this may be written

$$\underline{C}_t(t') = \underline{C}_o(-(t-t')) . \tag{4.82}$$

In order to obtain a frame-indifferent definition we require to generalise (4.82) to the condition that we can find an orthogonal tensor $\underline{Q}(t)$, satisfying $\underline{Q}(0) = \underline{I}$, such that

$$\underline{C}_t(t') = \underline{Q}(t)\cdot\underline{C}_o(-(t-t'))\cdot\underline{Q}^T(t) . \tag{4.83}$$

We take (4.83) as the basic definition of a motion with constant stretch history (or constant principal relative stretch history [T25]). These ideas were introduced by Coleman [C32,C33] who used the name "substantially stagnant motion", and Noll [N10] later the same year, introduced the name "motion with constant stretch history" (which we shall abbreviate as MCSH).

The idea of the MCSH also appears in the work of White [W11] who used the term "isoelastic flow" for a flow where the recoverable strain tensor is constant along a streamline and concluded that this requires the kinematics to be the same and hence that we have an MCSH. Olroyd [O3] introduced the idea of a "perpetual rheological history" in terms of convected coordinates and the equivalence of this and the MCSH has been proved [B4]. Lodge [L17] discussed the MCSH using body tensor fields and obtains results corresponding to those obtained using the space tensor fields we use here. An important point which Lodge makes is that an MCSH is not necessarily a rheologically steady flow since it includes at least one shear flow (orthogonal rheometer flow) in which the lines of shear rotate relative to the material [L17, p.137]. This leads Lodge to be "tempted to conjecture that no universally applicable, rheologically significant, meaning can be given to the term steady flow." This does not mean that the MCSH is irrelevant, but merely that it includes some unexpected flows because of the generalisation from (4.82) (which is too narrow a definition) to (4.83) (which appears to be broader than we might have hoped).

Nevertheless there are a number of useful results for the MCSH, starting with Noll [N10] who proved that a motion satisfies (4.83) if and only if

$$\underline{F}_o(-s) = \underline{Q}(-s)\cdot\exp(-s\underline{M}) \tag{4.84}$$

where $\underline{M}$ is a constant tensor. From (4.84), which uses a fixed reference configuration at time t = 0 we may obtain

$$\underline{F}_t(t') = \underline{Q}(t')\cdot\exp\left(-(t-t')\underline{M}\right)\cdot\underline{Q}^T(t) . \tag{4.85}$$

An exhaustive study of MCSH by Huilgol [H16,H17] follows the important theorem of Wang [W5] that (4.83) implies that $\underline{p}(t)$ is an isotropic function of the first three Rivlin-Ericksen tensors

$$\underline{A}_1(t) = 2\underline{e}(t), \quad \underline{A}_2(t) = \overset{\nabla}{\underline{A}}_1(t), \quad \underline{A}_3(t) = \overset{\nabla}{\underline{A}}_2(t) . \tag{4.86}$$

This leads to two classifications of all possible MCSH. First, on the basis of the properties of $\underline{M}$, we have (I) <u>viscometric flows</u> where $\underline{M}^2 = \underline{0}$, (II) <u>MCSH-II</u> where $\underline{M}^2 \neq \underline{0}$, $\underline{M}^3 = \underline{0}$ and (III) <u>MCSH-III</u> where $\underline{M}^n \neq \underline{0}$ for all n (this includes orthogonal rheometer flow and steady elongational flows) [N10,H16, Z1]. Secondly, on the basis of the properties of $\underline{A}_1$ (and $\underline{A}_2$) we have (a) $\underline{A}_1$ has distinct eigenvalues; then $\underline{A}_1$ and $\underline{A}_2$ suffice to determine $\underline{p}$ (this includes viscometric flows), (b) two eigenvalues of $\underline{A}_1$ are equal and $\underline{A}_1^2 = \underline{A}_2$; then $\underline{A}_1$ suffices to determine $\underline{p}$ (these flows are steady elongational flows) and (c) two eigenvalues of $\underline{A}_1$ are equal and $\underline{A}_1^2 \neq \underline{A}_2$.

This latter classification leads to a stronger result [H16,H15] that "a velocity field is an MCSH of simple extensional type if and only if $\underline{A}_1^2(t) = \underline{A}_2(t)$ for all t". A "simple extensional flow" is what we have called steady rectilinear extension. Note that the condition $\underline{A}_1^2(t) = \underline{A}_2(t)$ implies that the motion is an MCSH and also that the equal eigenvalues condition of (b) above is not required, so that steady elongations may be of class (a) or (b), as well as necessarily being MCSH-III. We may use these ideas to indicate how result (4.33) follows from (4.32) for steady rectilinear extension. It is clear that (4.32), $U_i = e_i X_i$, implies that $\underline{A}_2 = \underline{A}_1^2$ and hence the extra-stress is an isotropic function of $\underline{A}_1$. Then (4.33) follows immediately from a representation theorem for isotropic functions (in effect the Cayley-Hamilton theorem). The proof of (4.33) in the original paper [C31] is more direct.

5 Theoretical analysis of stretching

"Their language and the derivations of their language - religion, letters, metaphysics - all presuppose idealism."

Jorge Luis Borges

In this chapter we discuss a number of aspects of homogeneous uniaxial stretching for a few constitutive equations. The basic ideas have been set out in section 2.3 above, and experimental results for polymers have been presented in section 3.1. The motivation for this work is the desire to investigate the qualitative properties of constitutive equations, to find which of the predictions that they make correspond to expected, possible, unlikely, or completely ridiculous physical behaviour. This leads to an understanding of the roles played by the various parameters, and to limitations on the ranges of values which they may take (or which we should prefer them to take). The problem we encounter frequently is that it is only for the simplest equations, which have very limited quantitative predictive ability, that we can make strong or useful general statements. In the detailed description or prediction of experimental observations we are almost inevitably forced to compute numerical solutions to our equations. However an understanding of the qualitative behaviour of solutions, even for simplified models can help to organise and understand the computation. Interpretation of our mathematical results is generally straightforward, and here we concentrate on exhibiting the mathematical techniques and presenting some selected results.

Since some simple models have already been discussed in section 2.3, we start with a detailed consideration of the behaviour of the Jeffreys liquid, which has four parameters (material constants) in our formulation. We shall need to consider carefully the question of the appropriate initial conditions for the problem. Then we shall discuss principally two basic experiments; first stressing and relaxation and then creep and recovery. A similarly full discussion for the Maxwell model (as in [P16]) would not add much, and we confine remarks here to some significant differences in behaviour and to some points where we can make more progress in our qualitative studies.

The final section of this chapter reviews a number of studies of other constitutive equations in uniaxial stretching, and some studies of biaxial stretching. In particular we refer to comparisons with good experimental data, and to the effect of using a spectrum of relaxation times instead of the single relaxation time of our detailed studies here. In addition we discuss the effect of surface tension, and mention work on the stability of uniform stretching.

5.1 STRETCHING THE JEFFREYS LIQUID

5.1.1 Mathematical statement of the problem

As in section 2.3 we ignore gravity, inertia, surface tension and assume that we have a spatially homogeneous flow (2.64) and constant material parameters. The constitutive equation is (2.23), and this can be expressed in an integral form like (2.7) as

$$\underline{p}(t) = \int_{-\infty}^{t} \frac{2G}{\lambda}\left(1 - \frac{\Lambda}{\lambda}\right) e^{-(t-t')/\lambda} \underline{S}_t(t')dt' + 2\Lambda G\underline{e}(t) \tag{5.1}$$

where, as in the Maxwell model (2.27),

$$\underline{S}_t(t') = \frac{1}{2a}\begin{pmatrix} \{L(t)/L(t')\}^{2a}-1 & 0 & 0 \\ 0 & \{L(t)/L(t')\}^{-a}-1 & 0 \\ 0 & 0 & \{L(t)/L(t')\}^{-a}-1 \end{pmatrix} \tag{5.2}$$

because we are dealing with an elongational flow, and $G = \eta/\lambda$ as usual. The equivalence of the integral and rate equation forms (5.1) and (2.23) can be verified by differentiating (5.1), giving

$$\dot{\underline{p}}(t) = \int_{-\infty}^{t} \frac{2G}{\lambda}\left(1 - \frac{\Lambda}{\lambda}\right) e^{-(t-t')/\lambda}\left\{-\frac{1}{\lambda}\underline{S}_t(t') + \underline{e}(t)\cdot(2a\underline{S}_t(t') + \underline{I})\right\}dt' + 2\Lambda G\dot{\underline{e}}(t) \tag{5.3}$$

and then eliminating the integral involving $\underline{S}_t(t')$ between (5.1) and (5.3). We have used the fact that $\underline{S}_t(t) = \underline{0}$ and that for the flow given by (2.64)

$$d\underline{S}_t(t')/dt = \begin{pmatrix} \dot{\gamma}(t)\{L(t)/L(t')\}^{2a} & 0 & 0 \\ 0 & -\tfrac{1}{2}\dot{\gamma}(t)\{L(t)/L(t')\}^{-a} & 0 \\ 0 & 0 & -\tfrac{1}{2}\dot{\gamma}(t)\{L(t)/L(t')\}^{-a} \end{pmatrix} \tag{5.4}$$

where $\dot{\gamma}(t) = L(t)^{-1}dL(t)/dt$. The final result is

$$\underline{p} + \lambda(\dot{\underline{p}} - 2a\underline{e}\cdot\underline{p}) = 2\lambda G\{\underline{e} + \Lambda(\dot{\underline{e}} - 2a\underline{e}\cdot\underline{e})\} \tag{5.5}$$

which is the same as (2.23) for an irrotational flow with diagonal tensors $\underline{e}$ and $\underline{p}$ (both evaluated at the present time, t). The derivatives $\dot{\underline{p}}$ and $\dot{\underline{e}}$ are taken following a material particle, and for the spatially homogeneous flow considered in this chapter these are just ordinary derivatives of functions of t.

The axial stress, T, is the difference in axial and radial components of extra-stress, $p_{11} - p_{22}$, as before (2.79), and we may write this either from (5.1) as an integral,

$$T(t) = \int_{-\infty}^{t} \frac{G}{a\lambda}\left(1 - \frac{\Lambda}{\lambda}\right)e^{-(t-t')/\lambda}\left\{\left(\frac{L(t)}{L(t')}\right)^{2a} - \left(\frac{L(t)}{L(t')}\right)^{-a}\right\}dt' + 3\Lambda G\dot{\gamma}(t) \tag{5.6}$$

or (by eliminating p_{11} and p_{22} from the two independent component equations of (5.5)) as a differential equation,

$$\lambda^2\ddot{T} + (2-a\lambda\dot{\gamma}-\lambda\ddot{\gamma}/\dot{\gamma})\lambda\dot{T} + (1-a\lambda\dot{\gamma}-2a^2\lambda^2\dot{\gamma}^2-\lambda\ddot{\gamma}/\dot{\gamma})T$$

$$= 3\lambda G\{\dot{\gamma} + \Lambda(\ddot{\gamma}-a\dot{\gamma}^2) + \lambda\Lambda(\dddot{\gamma}-\ddot{\gamma}^2/\dot{\gamma}-2a^2\dot{\gamma}^3-a\ddot{\gamma}\dot{\gamma})\} \tag{5.7}$$

If $a = 1$ and $\Lambda = 0$ this reduces to (2.118), and we note that $a = 0$ is a special case requiring separate treatment. The range $0 < a \leq 1$ is of most interest, since this gives the best agreement with observed behaviour in shear and elongation for a number of polymers. We shall also require $G > 0$ and $0 \leq \Lambda \leq \lambda$, and note that the equality $\Lambda = \lambda$ gives us the Newtonian response, as (5.6) shows immediately.

5.1.2 Initial conditions

If the kinematics of the flow (i.e. the function of time, L(t)) are given then we have a second order differential equation for the stress T(t) with solution given by the integral (5.6). To calculate T(t) we require either L(t) for all past time t', or else initial conditions which will contain whatever information on the past flow history is necessary. From (5.7) we see that values of T and $\dot{T}$ at $t = t_o$ are needed, and these may be obtained from the integral equation (5.6) and its derivative. It is perhaps clearer that this is correct if we think of obtaining p_{11} and p_{22} from (5.1) and thence $\dot{p}_{11}$ and $\dot{p}_{22}$ from (5.5). By writing the integral (5.6) in the form

$$T(t) = \int_{-\infty}^{t_o} \frac{G}{a\lambda}\left(1 - \frac{\Lambda}{\lambda}\right) e^{-(t-t')/\lambda} \left\{\left(\frac{L(t)}{L(t')}\right)^{2a} - \left(\frac{L(t)}{L(t')}\right)^{-a}\right\} dt'$$

$$+ \int_{t_o}^{t} \frac{G}{a\lambda}\left(1 - \frac{\Lambda}{\lambda}\right) e^{-(t-t')/\lambda} \left\{\left(\frac{L(t)}{L(t')}\right)^{2a} - \left(\frac{L(t)}{L(t')}\right)^{-a}\right\} dt' + 3\Lambda G\dot{\gamma}(t) \tag{5.8}$$

we see that only the first integral involves knowledge of the flow for $t' < t_o$. This first integral may be written

$$\left(\frac{L(t)}{L(t_o)}\right)^{2a} e^{-(t-t_o)/\lambda} \int_{-\infty}^{t_o} \frac{G}{a\lambda}\left(1 - \frac{\Lambda}{\lambda}\right) e^{-(t_o-t')/\lambda} \left(\frac{L(t_o)}{L(t')}\right)^{2a} dt'$$

$$- \left(\frac{L(t)}{L(t_o)}\right)^{-a} e^{-(t-t_o)/\lambda} \int_{-\infty}^{t_o} \frac{G}{a\lambda}\left(1 - \frac{\Lambda}{\lambda}\right) e^{-(t_o-t')/\lambda} \left(\frac{L(t_o)}{L(t')}\right)^{-a} dt' \tag{5.9}$$

and use of (5.1) and (5.2) gives

$$\int_{-\infty}^{t_o} \frac{G}{a\lambda}\left(1 - \frac{\Lambda}{\lambda}\right) e^{-(t_o-t')/\lambda} \left(\frac{L(t_o)}{L(t')}\right)^{2a} dt' = p_{11}(t_o) - 2\Lambda G\dot{\gamma}(t_o) + \frac{G}{a}\left(1 - \frac{\Lambda}{\lambda}\right) \tag{5.10a}$$

and

$$\int_{-\infty}^{t} \frac{G}{a\lambda}\left(1 - \frac{\Lambda}{\lambda}\right) e^{-(t_o-t')/\lambda} \left(\frac{L(t_o)}{L(t')}\right)^{-a} dt' = p_{22}(t_o) + \Lambda G\dot{\gamma}(t_o) + \frac{G}{a}\left(1 - \frac{\Lambda}{\lambda}\right). \tag{5.10b}$$

Hence

$$T(t) = e^{-(t-t_o)/\lambda} \left\{\left(\frac{L(t)}{L(t_o)}\right)^{2a} \left(p_{11}(t_o) - 2\Lambda G\dot{\gamma}(t_o) + \frac{G}{a}\left(1 - \frac{\Lambda}{\lambda}\right)\right)\right.$$

$$\left. - \left(\frac{L(t)}{L(t_o)}\right)^{-a} \left(p_{22}(t_o) + \Lambda G\dot{\gamma}(t_o) + \frac{G}{a}\left(1 - \frac{\Lambda}{\lambda}\right)\right)\right\} + 3\Lambda G\dot{\gamma}(t) \tag{5.11}$$

$$+ \int_{t_o}^{t} \frac{G}{a\lambda}\left(1 - \frac{\Lambda}{\lambda}\right) e^{-(t-t')/\lambda} \left\{\left(\frac{L(t)}{L(t')}\right)^{2a} - \left(\frac{L(t)}{L(t')}\right)^{-a}\right\} dt' .$$

A generalisation of this result (for the upper convected Maxwell model) is established elsewhere [A16,P18] and discussed in connection with spinning. It makes explicit the formal correspondence between initial value problems for integral and rate equations.

Notice that both integral and rate equations for the Jeffreys model require $\dot{\gamma}(t)$ and, if we make the hypothesis that homogeneous stresses in a material should be bounded, then $\dot{\gamma}(t)$ must be defined, and bounded, for all t (and so L(t) must be differentiable). The rate equation requires additionally that $\dot{\gamma}(t)$ is twice differentiable and in this respect the integral equation is more general (unless we resort to the use of generalised functions in the rate equation). If there is a jump in $\dot{\gamma}(t)$ at t_o we see from (5.6) that there will be a corresponding jump in T(t),

$$T(t_o+) - T(t_o-) = 3\Lambda G[\dot{\gamma}(t_o+) - \dot{\gamma}(t_o-)]. \tag{5.12}$$

Here we use the notation $f(t_o+)$ and $f(t_o-)$ for the limiting values of a function f(t) as $t \to t_o$ with, respectively, $t > t_o$ and $t < t_o$, and use the fact that the integral in (5.6) is a continuous function of t.

This gives us an initial value of T(t) to be used in the differential equation (5.7) after a jump in $\dot{\gamma}(t)$, and we see that in addition we must find an initial value for $\dot{T}(t)$. To do this we differentiate (5.6) (or use (5.3)) at a time t for which $\dot{\gamma}(t)$ is differentiable, and hence so is T(t). This gives us

$$\begin{aligned} \dot{T}(t) + T(t) &= 3\Lambda G(\dot{\gamma}(t) + \lambda\ddot{\gamma}(t)) \\ &+ \int_{-\infty}^{t} G\left(1 - \frac{\Lambda}{\lambda}\right)e^{-(t-t')/\lambda}\dot{\gamma}(t)\left(2\left(\frac{L(t)}{L(t')}\right)^{2a} + \left(\frac{L(t)}{L(t')}\right)^{-a}\right)dt' . \end{aligned} \tag{5.13a}$$

If there is a jump in $\dot{\gamma}(t)$ at t_o, we obtain $T(t_o+)$ from (5.12) then take the limit at $t \to t_o+$ of (5.13) to get $\dot{T}(t_o+)$, and thus have the two necessary initial conditions if we know the history of the motion, L(t') for $-\infty < t' \leq t_o$. If we have a history in which $L(t) = L_o$ and $T(t) = 0$ for all $t < 0$ we may rewrite (5.13a) in the form

$$\begin{aligned} \lambda\dot{T}(t) + T(t) &= 3\lambda G\left(\dot{\gamma}(t)\left(\frac{\Lambda}{\lambda} + \left(1 - \frac{\Lambda}{\lambda}\right)\left(\frac{2}{3}\left(\frac{L(t)}{L_o}\right)^{2a} + \frac{1}{3}\left(\frac{L(t)}{L_o}\right)^{-a}\right)e^{-t/\lambda}\right) + \Lambda\ddot{\gamma}(t)\right) \\ &+ G\dot{\gamma}(t)\left(1 - \frac{\Lambda}{\lambda}\right)\int_{o}^{t} e^{-(t-t')/\lambda}\left(2\left(\frac{L(t)}{L(t')}\right)^{2a} + \left(\frac{L(t)}{L(t')}\right)^{-a}\right)dt' . \end{aligned} \tag{5.13b}$$

Then for $t \to 0+$ we have $L(t) \to L_o$ and the integral tends to zero, so that

$$\lambda\dot{T}(0+) + T(0+) = 3\lambda G[\dot{\gamma}(0+) + \Lambda\ddot{\gamma}(0+)]. \qquad (5.13c)$$

Thus the start of motion after a long period of rest in the stress-free state is described by the equations of linear viscoelasticity even though the stresses and rates of strain may be large. The fact that the strain is small is enough to ensure this, in contrast to the response of the Maxwell model where there is an instantaneous jump in strain which is nonlinearly related to the stress.

The alternative problem that we may pose, instead of the determination of the stress given the motion, is the determination of the motion given the stress. This includes the traditional creep test - elongation at constant stress. Elongation at constant force poses similar problems, though here instead of specifying $T(t)$ or $L(t)$ we specify a relation between them. In both cases we have a third order differential equation (5.7) for $L(t)$ but when the stress is specified the variable $L(t)$ does not appear explicitly in the equation (5.7) so that we may treat it as a second order equation for $\dot{\gamma}(t)$. Initial conditions may in principle be obtained from the history of the stress (by solving the appropriate integral equations for $L(t)$ and its derivatives). Here we consider only the situation where we know the solution for time $t < t_o$ and wish to deal with a jump in $T(t)$ at t_o (due to the sudden application or removal of a load). From (5.12) we obtain the jump in $\dot{\gamma}(t)$ corresponding to the given jump in $T(t)$, and hence, given $\dot{T}(t_o+)$, we may find $\ddot{\gamma}(t_o+)$ from (5.13) by taking the limit $t \to t_o+$. Note that there is no jump in $L(t)$; we may define $L(t)$ by

$$L(t) = L(t_1) \exp \int_{t_1}^{t} \dot{\gamma}(t')dt' \qquad (5.14)$$

and hence clearly a jump in $\dot{\gamma}$ just corresponds to a change in slope of the graph of $L(t)$. Alternatively we may argue that if there were a jump in $L(t)$, $\dot{\gamma}(t)$ would be undefined (or unbounded if we use generalised functions) and hence so would $T(t)$ from (5.6).

This approach applies equally well to constant force stretching, and in fact the situation is much simpler here than in the case of the Maxwell model (section 5.2.1) where a jump in $L(t)$ is possible. One point concerning jump conditions is that in practice the sudden application of a load is not truly instantaneous, and inertia (which we have neglected) will not allow an instantaneous change in velocity. Hence we may adopt an alternative approach to these idealisations, useful though they are, of assuming that $T(t)$, $L(t)$ and $\dot{\gamma}(t)$ (and possibly $\dot{T}(t)$ and $\ddot{\gamma}(t)$) are continuous functions. We then model the application of a stress (or a rate of strain) by a rapid but continuous rise over a short time interval. It is then necessary (or at least useful) to show that the precise form of the function describing the rise is immaterial at times which are large compared with the time of rise. This is a sort of temporal St. Venant's principle, and examples of this approach are given by Ferry [F6, pp.9-11] and in section 5.2.4 below.

5.1.3 Stressing and relaxation

This experiment (section 2.1.1, see also [G10] and [M24]) involves the stretching of a specimen at constant rate of strain between times t_o and t_1, and holding it at rest in the deformed (stretched) state for $t > t_1$. We assume that the specimen has been at rest and is unstressed for $t < t_o$, and for convenience we take $t_o = 0$. Hence for $0 < t < t_1$ we have to solve

$$\lambda^2\ddot{T} + (2-a\lambda\dot{\gamma})\lambda\dot{T} + (1-a\lambda\dot{\gamma}-2a^2\lambda^2\dot{\gamma}^2)T = 3\lambda G\dot{\gamma}\{1-a\Lambda\dot{\gamma}-2a^2\lambda\Lambda\dot{\gamma}^2\} \qquad (5.15)$$

with initial conditions

$$T(0+) = 3\Lambda G\dot{\gamma} \qquad (5.16a)$$

and

$$\dot{T}(0+) = 3G(1-\Lambda/\lambda)\dot{\gamma}\ . \qquad (5.16b)$$

We write $r = a\lambda\dot{\gamma}$, $\beta = \Lambda/\lambda$ and $E = 3G/a$ for the combinations of constants in the equation and have the solution

$$\frac{T(t)}{E} = \frac{r(1-\beta r-2\beta r^2)}{(1-2r)(1+r)} - \frac{2r(1-\beta)}{3(1-2r)}\,e^{-(1-2r)t/\lambda} - \frac{r(1-\beta)}{3(1+r)}\,e^{-(1+r)t/\lambda} \qquad (5.17a)$$

$$= \beta r + \frac{(1-\beta)r}{3}\left[\frac{2(1-e^{-(1-2r)t/\lambda})}{1-2r} + \frac{1-e^{-(1+r)t/\lambda}}{1+r}\right]. \qquad (5.17b)$$

We see that, as with the Maxwell model, this predicts a steady state which will be attained as $t \to \infty$ if $1-2r > 0$ and $(1+r) > 0$, i.e. if $-1 < r < \frac{1}{2}$.

The relaxation part of the experiment is even more simple to deal with, since it corresponds to $\dot{\gamma} = 0$ (or $r = 0$) in the above solution, with different boundary conditions. We have

$$\lambda^2\ddot{T} + 2\lambda\dot{T} + T = 0 \tag{5.18}$$

and hence

$$T(t) = (A+Bt)e^{-t/\lambda} \tag{5.19}$$

with A and B determined from the values of T and $\dot{T}$ at $t = t_1$ (obtained from (5.17)) using (5.12) and (5.13) in the form

$$T(t_1+) = T(t_1-) - 3\Lambda G\dot{\gamma} = T(t_1-) - \beta rE \tag{5.20a}$$

and

$$\dot{T}(t_1+) = -T(t_1+)/\lambda, \tag{5.20b}$$

whence $B = 0$ and $A = T(t_1+)e^{t_1/\lambda}$.

The conclusion (which is not altogether unexpected) is that in the absence of motion the stresses decay (or relax) in exactly the same way as for a linear viscoelastic material. This will hold for a more realistic model involving a spectrum of relaxation times, when initial conditions will be needed for each partial stress. Observed behaviour, that initial stress relaxation is more rapid for larger strain rates cannot be accommodated in this way, for although the nonlinearity affects the initial conditions through the non-linearity of (5.17), this has the reverse effect, since the values of the partial stresses in steady flow are increased more for larger λ, which corresponds to slower relaxation.

5.1.4 Creep and recovery

In this experiment we apply a constant stress $T = T_o$ for $0 < t < t_1$, with $T = 0$ for $t < 0$ and $t > t_1$. We find that, if we write $x = a\lambda\dot{\gamma}$, we have a second order equation (obtained from (5.7)) for x as a function of t/λ,

$$\beta\{x'' - x'^2/x - x'x - 2x^3 + x' - x^2\} + x = K\{1 - x - 2x^2 - x'/x\} \tag{5.21}$$

where $\beta = \Lambda/\lambda$ as before, $K = T_o/E = aT_o/3G$ and $x' = \lambda dx/dt$. The initial conditions for this are

$$x(0+) = a\lambda\dot{\gamma}(0+) = a\lambda T_o/3\Lambda G = K/\beta \tag{5.22a}$$

and

$$x'(0+) = a\lambda^2\ddot{\gamma}(0+) = aT_o\lambda(\Lambda-\lambda)/3\Lambda^2 G = K(\beta-1)/\beta^2 . \tag{5.22b}$$

An elongation at constant rate corresponds to a constant value of x and, while the stressing experiment gives a unique constant stress for any value of r (or x) in the interval $(-1, \frac{1}{2})$, we find here a cubic for the steady rate of strain,

$$f(x) \equiv 2\beta x^3 + (\beta - 2K)x^2 - (1+K)x + K = 0. \tag{5.23}$$

This has three real roots for $0 < \beta < 1$, two of which are positive, and examination of the sign of $f(\frac{1}{2})$, $f(K)$, $f(K/\beta)$ shows that one root is less than $\frac{1}{2}$ and less than K, while the other is greater than $\frac{1}{2}$ and greater than K/β.

There is, no doubt, more that a study of (5.21) will reveal. Here we pick out one simple consequence of (5.12), that when the stress is removed there is a reduction in $\dot{\gamma}$ by $T_o/3G\Lambda$. Hence if we apply T_o for long enough for steady flow to be attained, assuming this is possible, we have

$$\dot{\gamma}(t_1+) = \dot{\gamma}_s - T_o/3G\Lambda \tag{5.24a}$$

or

$$a\lambda\dot{\gamma}(t_1+) = x_s - K/\beta \tag{5.24b}$$

where $\dot{\gamma}_s$ corresponds to the root x_s of (5.23). If x_s is the smaller positive root of (5.23) we find $\dot{\gamma}(t_1+) < 0$, i.e. the material recovers or retracts when the stress is removed, which is what we expect. However if x_s is the larger root, $\dot{\gamma}(t_1+) > 0$, i.e. the material continues to stretch when the stress has been removed. This is not physically likely, and in fact a result due to Lodge, McLeod and Nohel [L18] proves that the physically expected result is obtained for a = 1 in recovery after stretching at constant rate (which is a slightly different situation) (see section 5.1.5).

The recovery follows the equation (5.21) with K = 0 for $t > t_1$, with initial conditions (5.24) (or the equivalent with $\dot{\gamma}(t_1-)$ replacing $\dot{\gamma}_s$ if steady flow has not been attained by $t = t_1$) and

$$\lambda\ddot{\gamma}(t_1+) = -\dot{\gamma}(t_1+)\left\{1 + (1-\Lambda/\lambda)/\Lambda\int_{-\infty}^{t_1} e^{-(t_1-t')/\lambda}\left[\frac{2}{3}\left(\frac{L(t_1)}{L(t')}\right)^{2a} + \frac{1}{3}\left(\frac{L(t_1)}{L(t')}\right)^{-a}\right]dt'\right\}. \tag{5.25}$$

It is not easy to obtain estimates of the recoverable strain from this so one of the main questions which was raised by the simpler Maxwell model analysis is unresolved. This refers to the discrepancy between observed measurements of recoverable strain, which show that it increases with t_1, and the Maxwell model predictions (sections 2.3.7 and 5.2.3 below) with recoverable strain starting at its maximum value and decreasing.

An indication that the Jeffreys model predicts more realistic behaviour is given by very simple approximation for small t_1 illustrated in Figure 5.1. We approximate $\dot{\gamma}$ by a linear function of t, so that

$$\dot{\gamma}(t) \simeq (T_o/3\Lambda G)(1 - (1-\beta)t/\Lambda) \tag{5.26}$$

for $0 < t < t_1$. We shall use

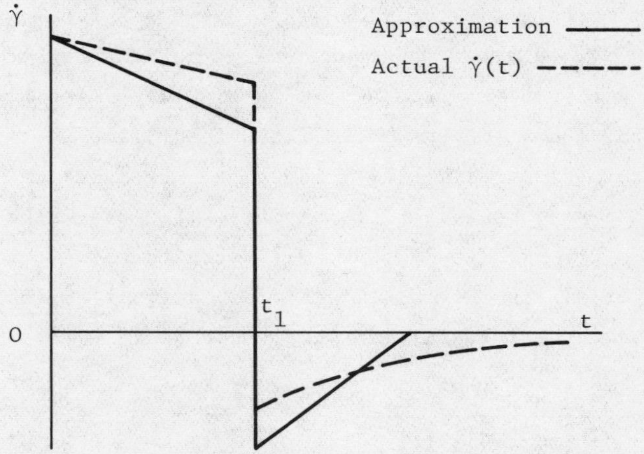

Figure 5.1 *Creep and recovery approximations for the Jeffreys liquid.*

$$\dot{\gamma}(t) \simeq \dot{\gamma}(t_1+) + (t - t_1)\ddot{\gamma}(t_1+) \tag{5.27}$$

for $t > t_1$ in order to get an estimate (probably an underestimate) of the recovery. From (5.12) and (5.26) we obtain

$$\dot{\gamma}(t_1+) \simeq -(1-\beta)T_o t_1/3\Lambda^2 G \tag{5.28}$$

and from (5.13b), ignoring $O(t_1)$ terms we find

$$\lambda\ddot{\gamma}(t_1+) \simeq -\dot{\gamma}(t_1+)\ (1 + (1-\beta)/\beta) \tag{5.29a}$$

$$= (1-\beta)T_o t_1/3\Lambda^2 G\beta. \tag{5.29b}$$

Integrating (5.26) and (5.27) gives us the strains

$$\gamma(t_1) \equiv \log\{L(t_1)/L(0)\} \simeq T_o t_1/3\Lambda G \tag{5.30}$$

and

$$\gamma_r(t_1) \equiv \log\{L(t_1)/L(\infty)\} \simeq (1-\beta)T_o t_1/6\Lambda G \tag{5.31a}$$
$$= (1-\beta)\gamma(t_1)/2 .$$

(The integral used for γ_r is $-\int_{t_1}^{t_2} \dot{\gamma}(t)dt$, where t_2 is the time at which $\dot{\gamma}(t)$, given by (5.27), becomes zero. Hence $\gamma_r = \frac{1}{2}[\dot{\gamma}(t_1+)]^2/\ddot{\gamma}(t_1+)$).

We see then that the initial response of the material has $0 < \gamma_r < \gamma$, which is physically realistic, provided that $0 < \beta < 1$ (and $\gamma_r = 0$ for $\beta = 1$, when the constitutive equation and equations of motion are satisfied by a solution of the same problem for a Newtonian liquid). Also both γ and γ_r increase from zero, and if we can prove that steady flow is attained (i.e. that $\dot{\gamma}(t)$ tends to a finite limiting value as t increases) we have a realistic model of the qualitative behaviour of polymeric liquids in the creep and relaxation experiment. If steady flow, with stretch rate $\dot{\gamma}_s$ is attained, and then maintained for a long enough time we find, as well as (5.24) for $\dot{\gamma}(t_1+)$, that we may evaluate the integral in (5.25) and obtain

$$\ddot{\gamma}(t_1+) = \frac{T_o(1-\beta)}{3G\Lambda^2(1-2a\lambda\dot{\gamma}_s)(1+a\lambda\dot{\gamma}_s)} \tag{5.32a}$$

$$= \frac{T_o(T_o-3\beta G)}{(3\Lambda G)^2\lambda\dot{\gamma}_s} . \tag{5.32b}$$

It is easy to express this in terms of $\dot{\gamma}_s$, but not to do so explicitly in terms of the given value of T_o, since these are related by the limiting value of (5.17) as $t \to \infty$, namely

$$\frac{T_o}{3G} = \frac{\lambda\dot{\gamma}(1-\beta a\lambda\dot{\gamma}-2\beta a^2\lambda^2\dot{\gamma}^2)}{(1-2a\lambda\dot{\gamma})(1+a\lambda\dot{\gamma})} \tag{5.33}$$

We observe that, from (5.32a) provided that $\beta < 1$ and $a\lambda\dot{\gamma} < \frac{1}{2}$, $\ddot{\gamma}(t_1+) > 0$. The proof of the expected result that $\dot{\gamma}(t_1+) < 0$ has not been accomplished yet from a consideration of the differential equation (5.21).

5.1.5 Existence and properties of solutions

An important series of results have been obtained by Lodge, McLeod and Nohel

[L18] on the existence and properties of solutions of a generalisation of the integro-differential equation (5.6) for a flow similar to those we have been discussing. The flow considered is stretching at constant rate of strain (the stressing experiment) for times $-t_o < t < 0$ (with the material at rest for $t < -t_o$) followed by recovery (stress reduced to zero for $t > 0$, "unconstrained" or free recovery as in the flow considered in the previous sections for $t > t_1$ following creep). In addition to discussing this problem (the flow for $t > 0$) the limit of $\Lambda \to 0+$ is discussed, leading to the idea of the Jeffreys liquid as a singular perturbation of a Maxwell liquid. We shall not discuss the generalisations in detail here; they concern both the memory function and the deformation tensor in the constitutive equation. The memory function is a general function of (t-t') with properties motivated by those of the discrete spectrum of relaxation times (2.26b). The deformation tensor has properties of the Finger deformation tensor of the rubber-like liquid (a = 1 in the model we have considered here), and the generalisation at present excludes the deformation tensor (5.2) if a < 1.

For $T(t) = 0$ $(t > 0)$, and $a = 1$, (5.6) gives

$$-3\Lambda G\dot{\gamma}(t) \quad = \quad \int_{-\infty}^{t} \frac{G}{a\lambda}\left(1 - \frac{\Lambda}{\lambda}\right) e^{-(t-t')/\lambda} \left\{ \left(\frac{L(t)}{L(t')}\right)^{2} - \left(\frac{L(t)}{L(t')}\right)^{-1} \right\} dt'. \qquad (5.34)$$

If we write $\dot{\gamma}(t) = \dot{L}(t)/L(t)$, $\mu = 3\Lambda G$,

$$m(t-t') \quad = \quad (G/a\lambda)(1-\Lambda/\lambda)\exp\{-(t-t')/\lambda\} \qquad (5.35a)$$

and define

$$F\{y,z\} \quad = \quad y^{3}/z^{2} - z \qquad (5.35b)$$

we obtain essentially the Lodge equation (with a single relaxation time instead of a spectrum.

$$-\mu\dot{L}(t) \quad = \quad \int_{-\infty}^{t} m(t-t')F\{L(t), L(t')\}dt'. \qquad (5.36)$$

This is to be solved for $t > 0$, when

$$L(t) \quad = \quad g(t) \quad = \quad \begin{cases} 1 & \text{for } -\infty < t \leq t_o \\ \exp\{\dot{\gamma}(t+t_o)\} & \text{for } -t_o < t \leq 0 \end{cases} \qquad (5.37)$$

for a given constant stretch rate, $\dot{\gamma}$. A more general problem replaces the specified g(t) by a given function with a minimum number of properties. These are (for most of Lodge's results) that $g(-\infty) = 1$, $g(0) > 1$ and g(t) is continuous and non-decreasing.

The first theorem [L18] states that for each $\mu > 0$ the initial value problem (5.36), (5.37) has a unique solution $L(t) = \phi(t;\mu)$ for $0 \leq t < \infty$ with the properties

$$\dot{\phi}(t;\mu) < 0 \tag{5.38a}$$

$$1 < \phi(t;\mu) \leq g(0) \tag{5.38b}$$

and if $\mu_1 > \mu_2$ then

$$\phi(t;\mu_1) > \phi(t;\mu_2). \tag{5.38c}$$

Furthermore if $g_1(t) \geq g_2(t)$, where both have the properties listed above, and the corresponding solutions are $\phi_1(t;\mu)$ and $\phi_2(t;\mu)$ then

$$\phi_1(t;\mu) \geq \phi_2(t;\mu). \tag{5.38d}$$

As a corollary the limit of $\phi(t;\mu)$ as $t \to \infty$ exists, and if we write

$$\lim_{t\to\infty} \phi(t;\mu) = \alpha(\mu) \tag{5.39a}$$

then

$$1 \leq \alpha(\mu) \leq g(0) \tag{5.39b}$$

and, if $\mu_1 > \mu_2$,

$$\alpha(\mu_1) \geq \alpha(\mu_2) . \tag{5.39c}$$

This inequality (5.38d) establishes that if we compare two specimens 1 and 2, and specimen 1 is stretched more than specimen 2 at all times before $t = 0$, length of specimen 1 at any subsequent time is greater than that of specimen 2 at the same time - even if specimen 1 retracts more rapidly it does not overtake specimen 2. Experimental evidence appears to suggest the contrary in shear, but is not available in elongation. In addition, the greater the parameter μ (or Λ in our previous notation - with G altered to keep $G(1-\Lambda/\lambda)$ unchanged) the greater the value of $\phi(t;\mu)$ and hence the smaller the recovery $g(0) - \phi(t;\mu)$, both in a comparison at any chosen time t and, from (5.39c), in comparison of the ultimate recovery $g(0) - \alpha(\mu)$. Hence μ plays the role of a damping coefficient for the recovery - consistent with our spring and dashpot model of Figure 2.1, where $\mu = G\Lambda$ is the viscosity of the dashpot in parallel with the Maxwell element. In models of polymer rheology the Maxwell elements represent the polymer molecule response and the viscous element represents either the solvent in dilute solution theory or low molecular mass polymer in the network theory for polymer melts.

There is not space to discuss fully the theorems of Lodge, McLeod and Nohel, which lead to upper and lower bounds for $\phi(t;\mu)$ and an investigation of the solution behaviour for $\mu \to 0$. When $\mu = 0$ we have the Maxwell model, and a boundary layer near $t = 0$ is found for $\mu \to 0+$. We mention here two results, an existence theorem based on weaker hypotheses on the generalised form of $F\{y,z\}$ in (5.36), and stronger results for our special case of a single relaxation time (5.35a). If $F\{y,z\}$ is continuous for positive values of its arguments and

$$F\{x,x\} = 0, \quad \text{for } x > 0, \tag{5.40a}$$

$$F\{y,z\} > 0 \text{ for } y > z > 0, \tag{5.40b}$$

$$F\{y,z\} < 0 \text{ for } 0 < y < z \tag{5.40c}$$

then there is at least one solution of (5.36) and (5.37) (and $1 < L(t) < g(0)$ for $t > 0$) for any $\mu > 0$. These hypotheses allow our equation (5.6) (with $T(t) = 0$ of course, but with any non-zero value of a). The single exponential memory function $m(t-t')$ gives us, in place of (5.38b),

$$1 < y_o < \phi(t;\mu) \leq g(0) \text{ for } t > 0 \tag{5.41}$$

(and of course $\alpha(\mu) \geq y_o$) where y_o is uniquely defined by

$$\int_{-\infty}^{o} e^{t'/\lambda} F\{y_o, g(t')\}dt' = 0 \tag{5.42a}$$

or, using (5.35b),

$$y_o^3 = \int_{-\infty}^{o} e^{t'/\lambda} g(t')dt' \Big/ \int_{-\infty}^{o} e^{t'/\lambda} g(t')^{-2} dt' . \tag{5.42b}$$

Now, as we shall see, this value of y_o is the instantaneous recovery for the Maxwell model ($\mu = 0$ and a single relaxation time). Moreover

$$\lim_{\mu \to 0+} \alpha(\mu) = y_o > 1 \tag{5.43}$$

when $\alpha(\mu)$ is the ultimate length of the specimen, defined in (5.39a). The proofs of these various results [L18] illustrate the power of integral equation techniques, and offer an interesting contrast to the differential equation studies here.

5.2 STRETCHING THE MAXWELL LIQUID

5.2.1 The mathematical model

The equations are just those obtained by setting $\Lambda = 0$ in those of section 5.1.1, namely

$$T(t) = \int_{-\infty}^{t} \frac{G}{a\lambda} e^{-(t-t')/\lambda} \left[\left(\frac{L(t)}{L(t')}\right)^{2a} - \left(\frac{L(t)}{L(t')}\right)^{-a} \right] dt' \qquad (5.44)$$

or, if the indicated differentiations are possible,

$$\lambda^2 \ddot{T} + (2-a\lambda\dot{\gamma}-\lambda\ddot{\gamma}/\dot{\gamma})\lambda\dot{T} + (1-a\lambda\dot{\gamma}-2a^2\lambda^2\dot{\gamma}^2-\lambda\ddot{\gamma}/\dot{\gamma})T = 3\lambda G\dot{\gamma} . \qquad (5.45)$$

The special case $a = 0$ is dealt with in section 2.3.6 for the rate equation (see equations (2.110), (2.111), (2.112)) and has been discussed elsewhere [P15]. The corresponding integral model is obtained by taking the limit $a \to 0$, and gives

$$T(t) = \int_{-\infty}^{t} \frac{3G}{\lambda} e^{-(t-t')/\lambda} \log\left(\frac{L(t)}{L(t')}\right) dt' . \qquad (5.46)$$

for this flow.

The fundamental difference between this model and the Jeffreys liquid is associated with their short-term response (see section 2.1.1). For the Maxwell liquid a jump in stress is associated not with a jump in strain rate but with a jump in strain or specimen length. If $L(t') = L(0) = 1$ for $t' < 0$ (the choice of length scale involves no loss of generality) then (5.44 becomes

$$T(t) = \frac{G}{a} e^{-t/\lambda}\{L(t)^{2a} - L(t)^{-a}\}$$
$$+ \int_{0}^{t} \frac{G}{a\lambda} e^{-(t-t')/\lambda} \left[\left(\frac{L(t)}{L(t')}\right)^{2a} - \left(\frac{L(t)}{L(t')}\right)^{-a} \right] dt' \qquad (5.47)$$

and hence, taking the limit as $t \to 0+$ and assuming that $L(t)$ is bounded,

$$T(0+) = \frac{G}{a} \{L(0+)^{2a} - L(0+)^{-a}\}. \qquad (5.48)$$

For (5.46) we obtain instead the equation

$$T(0+) = 3G \log (L(0+)), \tag{5.49}$$

and in both cases we can readily show that there is a one to one relationship between $T(0+)$ and $L(0+)$, for $0 \leq T(0+) < \infty$ and $1 \leq L(0+) < \infty$. The case of compression (equal biaxial stretching) can also be included, with $-\infty < T(0+) \leq 0$ corresponding to $0 < L(0+) \leq 1$.

The technique of differentiating (5.44) and then taking the limit allows us to obtain $\dot{\gamma}(0+)$ also, and hence the necessary initial conditions for the creep and recovery experiment, or for stretching at constant force. We shall not discuss the latter below, and make two observations here. Constant force stretching is exactly analogous to spinning if we make the transformation $\dot{\gamma}(t) \to U'(X)$, $L(t) \to U(X)$ and $d(\cdot)/dt = U(X)\, d(\cdot)/dX$. Hence we may use the general equations like (5.44) or (5.45) to obtain (or to check) spinning equations, and deductions about either flow will be relevant to the other. There are differences, notably that spinning is normally a two-point boundary-value problem rather than an initial-value problem, and of course we are neglecting inertia, gravity, surface tension here. We note also that the initial conditions for constant force stretching differ from those for constant stress stretching, since $T(0+)$ in (5.48) or (5.49) is replaced by $F/A(0+) = FL(0+)/A(0-)$. These altered equations may have no solution for $a < \frac{1}{2}$, corresponding to the fact that for large applied force we have a situation where the reduction in area and consequent increase in stress cannot be balanced by a further extension, however large. This non-existence of solutions to a simple problem in nonlinear elasticity has been observed before [R8]; the material will not support a load in excess of some critical amount.

5.2.2 Stressing and relaxation

Where the kinematics are prescribed and are possible for the Jeffreys liquid we may obtain the results for the Maxwell liquid just by setting $\Lambda = 0$, so that we have to solve

$$\lambda^2\ddot{T} + (2-a\lambda\dot{\gamma})\lambda\dot{T} + (1-a\lambda\dot{\gamma}-2a\lambda^2\dot{\gamma}^2)T = 3\lambda G\dot{\gamma} \tag{5.50}$$

subject to

$$T(0+) = 0, \quad \dot{T}(0+) = 3G\dot{\gamma} \tag{5.51a}$$

These correspond to conditions $\underline{p}(0+) = \underline{0}$ on the extra-stress tensor of the virgin material. The solution obtained is the well-known extension [E8,P16] of the result of Denn and Marrucci for the upper convected Maxwell model [D10],

$$T(t) = \frac{3\lambda G\dot{\gamma}}{(1-2a\lambda\dot{\gamma})(1+a\lambda\dot{\gamma})} - \frac{2\lambda G\dot{\gamma}}{(1-2a\lambda\dot{\gamma})} e^{-(1-2a\lambda\dot{\gamma})t/\lambda} - \frac{G\lambda\dot{\gamma}}{(1+a\lambda\dot{\gamma})} e^{-(1+a\lambda\dot{\gamma})t/\lambda} \tag{5.52}$$

with the steady state, for $-1 < a\lambda\dot{\gamma} < \frac{1}{2}$,

$$T = 3\lambda G\dot{\gamma}/(1-2a\lambda\dot{\gamma})(1+a\lambda\dot{\gamma}) . \tag{5.53}$$

On cessation of the motion at $t = t_1$ the stress relaxes exponentially, as for the linear model, and exactly as does the Jeffreys liquid (5.19)(5.20), but with no jump in stress at $t = t_1$.

5.2.3 Creep and recovery

Here we need (5.48) or (5.49) and, as indicated earlier, an equation for $\dot{\gamma}(0+)$, since we have a second order differential equation for $\dot{\gamma}(t)$ (or for $L(t)$), though we may deal first with (5.45) as a first order equation for $\dot{\gamma}(t)$. The appropriate condition is

$$\lambda\dot{\gamma}(0+) = T/\{G[2L(0+)^{2a} + L(0+)^{-a}]\}. \tag{5.54}$$

For $a = 0$, since $\dot{\gamma} = T/3\eta$ which is constant, this is not needed.

In order to make some deductions concerning the solutions of (5.45) for $T(t) = T_o$, constant, i.e. solutions of

$$\lambda\ddot{\gamma} = \dot{\gamma}\{1 - (a-3G/T_o)\lambda\dot{\gamma} - 2a\lambda^2\dot{\gamma}^2\} \tag{5.55}$$

it is useful to combine (5.48) and (5.54), eliminating $L(0+)$, to obtain

$$a\lambda\dot{\gamma}(0+) = (aT_o/3G)(1-2a\lambda\dot{\gamma}(0+))^{2/3}(1+a\lambda\dot{\gamma}(0+))^{1/3} \tag{5.56a}$$

or

$$(1-2a\lambda\dot{\gamma}(0+))^{2/3}(1+a\lambda\dot{\gamma}(0+))^{1/3} = 3G\lambda\dot{\gamma}(0+)/T_o . \tag{5.56b}$$

There is a unique positive root of this equation corresponding to the unique root $L(0+) > 1$, and this root lies in the interval $0 < \dot{\gamma}(0+) < 1/2a\lambda$. (For details of proof of this see [P16].) Then we notice that for $\dot{\gamma} > 0$ the right-hand side of (5.55) is a monotone decreasing function of $\dot{\gamma}$, and so if we can prove that this is negative for $\dot{\gamma} = \dot{\gamma}(0+)$ we can deduce that $\dot{\gamma}(0+) > \dot{\gamma}_s$, the steady stretch rate (which gives $\ddot{\gamma} = 0$). This proof is accomplished by using (5.56b) to eliminate T_o from (5.55), which becomes

$$\lambda\ddot{\gamma} = \dot{\gamma}(0+)\{(1-2a\lambda\dot{\gamma}(0+))(1+a\lambda\dot{\gamma}(0+)) - (1-2a\lambda\dot{\gamma}(0+))^{2/3}(1+a\lambda\dot{\gamma}(0+))^{1/3}\} \tag{5.57}$$

and for $0 < \dot{\gamma}(0+) < 1/2a\lambda$ this function is indeed negative.

For this model we have instantaneous recovery on removal of the stress and no delayed recovery at all (i.e. no further motion). The amount of the recovery is given, for the general case, where T and $\dot{T}$ denote values at t_1- and $T(t_1+) = 0$, by

$$\left(\frac{L(t_1+)}{L(t_1-)}\right)^{3a} = \frac{\lambda\dot{T} + (1-2a\lambda\dot{\gamma}(t_1-))T}{\lambda\dot{T} + (1+a\lambda\dot{\gamma}(t_1-))T} \tag{5.58}$$

For creep this gives, with $\dot{T} = 0$,

$$\begin{aligned}\gamma_r(t_1) &\equiv \log\{L(t_1-)/L(t_1+)\} \\ &= (1/3a)\log\{[1+a\lambda\dot{\gamma}(t_1-)]/[1-2a\lambda\dot{\gamma}(t_1-)]\}\end{aligned} \tag{5.59}$$

If $a = 1$ and we identify $\dot{\gamma}(t_1-)$ with the constant $\dot{\gamma}$ in (5.37) we see that γ_r is the same as y_o in (5.42) if t_o is large enough, i.e. if steady flow has been attained. If this is not the case, then in the constant stretch rate flow $\dot{T}(t)$ is non zero, and has to be determined to obtain γ_r from (5.58), or else we must evaluate the integrals in (5.42b).

The important fact about (5.59) is that the recovery after creep is a decreasing function of time, because $\dot{\gamma}(t)$ is a decreasing function. In the limit $t_1 \to 0$ we find $\gamma_r = \gamma(0+)$, that is to say that all the instantaneous deformation is recoverable initially, but part is dissipated in some way during the flow. This is not easily explained in terms of springs and dashpots, and is essentially a nonlinear effect.

5.2.4 Initiation of creep

As mentioned in section 5.1.2 above, there is the possibility of obtaining an analogue of St. Venant's principle, that the precise details of how a load is applied will not influence behaviour except for a short time after the application. It is in fact hard to prove any such result in general, and here we discuss the only successful attempt, on the simplest of models.

We take the co-rotational Maxwell model for which we have, in any uniaxial stretching flow

$$\lambda\dot{T}(t) + T(t) = 3\eta\dot{\gamma}(t). \tag{2.110}$$

Then if T(t) increases linearly from 0 at t = 0 to T_o at $t = \varepsilon$ we have, for $0 < t < \varepsilon$

$$T(t) = T_o t/\varepsilon \tag{5.60a}$$

and hence

$$\dot{\gamma}(t) = (\lambda+t)T_o/3\eta\varepsilon . \tag{5.60b}$$

If the specimen length L(t) is continuous, and L(0) = 1 then integrating (5.60b), using $\dot{L}(t)/L(t) = \dot{\gamma}(t)$, gives

$$\log\{L(t)\} = (\lambda t+\tfrac{1}{2}t^2)T_o/3\eta\varepsilon \tag{5.61a}$$

whence

$$L(\varepsilon) = \exp\{(\lambda T_o/3\eta)(L+\varepsilon/2\lambda)\}. \tag{5.61b}$$

For $t > \varepsilon$ we take $T(t) = T_o$, constant, and

$$L(t) = L(\varepsilon)\, \exp\{(\lambda T_o/3\eta)(t-\varepsilon)/\lambda\} \tag{5.62}$$

For $\varepsilon \to 0$ we recover equation (5.49), i.e. $L(0+) = \exp(T_o/3G)$ where $3G = 3\eta/\lambda$ which is the Young's modulus for the elastic response of this model.

If $\varepsilon \neq 0$ we have

$$L(t) = L_o \exp\{(T_o/3G)(t/\lambda)\}\exp\{-\varepsilon T_o/6G\lambda\} \tag{5.63}$$

and clearly for small ε the relative error in taking L(t) to be the value predicted for $\varepsilon = 0$ is $O(\varepsilon)$. Stronger results (not requiring the precise form of T(t) to be specified as in (5.60a)) can undoubtedly be proved, but there seems to be no great merit in this, since the model is such a simple (and unrealistic) one. We note that $\dot{\gamma}(t)$ is constant for $t > \varepsilon$, when T(t) is constant, and for $0 < t < \varepsilon$, as $\varepsilon \to 0$ we have $\dot{\gamma}(0+)$ unbounded (i.e. a jump in stress corresponds to a jump in strain and an undefined strain rate).

It is possible to try to examine the case of constant force stretching in this way, but this has not been carried to a clear conclusion to compare with the simple discussion of the jump conditions for the corotational models [P15]. A detailed study of this ought to include allowance for the fluid inertia, and we recall from section 4.3.1, equation (4.40) in particular, that uniform stretching is not possible without a non-uniform external pressure when inertia is significant. Thus there are awkward problems, and this does not seem to be an area where much effort should be directed.

5.3 REVIEW

5.3.1 Uniaxial stretching

There have been a number of comparisons of experimental results with predictions of more complicated constitutive equations. The steady flow predictions are given in section 2.2, where the start-up of stretching at constant rate of strain is briefly discussed. We mention here a few results which are recent or significant (or possibly both).

The use of a Maxwell model with a spectrum of relaxation times (2.26) [C8] is noticeably more successful than the single relaxation time model discussed above. The prediction of the model for the stressing and relaxation experiment is for the stress to be given by a sum of terms like (5.52). It is easy to make calculations for this model where the kinematics are prescribed, and the only noteworthy feature is that the critical strain rate, above which the stress does not tend to a finite limit as time increases (there is no steady state), is determined by the largest relaxation time, corresponding to the slowest mode of stress relaxation. This is discussed in a little more detail elsewhere [P16] and the point is made there that if this mode makes a small contribution to the spectrum the critical term in the stress will not become significant for a correspondingly large time (the smaller the modulus of the slowest mode the longer the time before the partial stress becomes significant). A similar point is made by Ting and Hunston [T19], in discussing difficulties in the experimental observations of the predicted high stresses.

Denn [D14] devotes a substantial part of a useful review to the behaviour of a Maxwell model with a spectrum. He expresses qualitative behaviour in terms of mean values of the parameters and obtains some interesting asymptotic results. We may note that definition of a mean relaxation time, modulus and viscosity may be done in various ways. Denn chooses to use mean values $\bar{\lambda}$, $\bar{G}$ and $\bar{\eta}$ (defined below, (6.52)) which are directly obtainable from conventional viscometric experiments [D13,D14]. This, of course, means that the mean values are appropriate to shear flow and as we have noted (and Denn notes) the longer relaxation times have a greater influence in elongational flows. This is discussed by the author at slightly greater length in connection with spinning [P18], but does not imply a major criticism of Denn's work. Of his results we note here that the initial rate of increase of stress in the stressing experiment is given by

$$\lim_{t\to 0} \dot{T}(t) = 3 \sum_{k=1}^{M} G_k \dot{\gamma} \geq 3\overline{G}\dot{\gamma} \tag{5.64}$$

with equality only for the case of a single relaxation time, M = 1. (The inequality follows from the definition of $\overline{G}$, (6.52) below, and an elementary extension of the Cauchy-Schwarz inequality.) Also for small strain rates ($\lambda_M \dot{\gamma} << 1$) the steady state is given by

$$\lim_{t\to\infty} T(t) = 3\overline{\eta}\ (1+\overline{\lambda}\dot{\gamma})\dot{\gamma} \tag{5.65}$$

which is the same as for a Maxwell model with a single relaxation time $\overline{\lambda}$ and modulus $\overline{G} \equiv \overline{\eta}/\overline{\lambda}$.

Deviations from the predictions of theory for the rubberlike liquid (upper convected Maxwell model) with a spectrum of five relaxation times [C8] are discussed by Lodge [L16,L17]. He defines an invariant of the rate of strain tensor in terms of principal rates of strain e_ℓ by

$$I_{IV} = 2^{-\frac{1}{2}} \left(\left(\sum_{\ell=1}^{3} e_\ell{}^2 \right)^{\frac{1}{2}} + \left| \sum_{\ell=1}^{3} e_\ell{}^3 \right|^{1/3} \right) \tag{5.66}$$

(so that $I_{IV} = 3.01\dot{\gamma}$ in uniaxial stretching and $I_{IV} = \kappa$ in simple shear). The ratio of the difference of principal stresses predicted to that observed experimentally was found to depend in the same way on $I_{IV}t$ for both shear and elongation, and in both cases disagreement between theory and experiment becomes significant at a value for $I_{IV}t$ of about 3. These results refer to the measurement and prediction of stress growth when flows with constant rate of strain are started for a low-density polyethylene melt.

More recent data on a very similar melt [L3] has been used by Wagner [W2] who fitted a model with a strain dependent relaxation spectrum (of the K-BKZ type (2.38), discussed briefly in section 2.2.4). The equation is

$$\underline{p}(t) = \int_{-\infty}^{t} m(t-t')h(t,t')\underline{C}_t^{-1}(t')dt' \tag{5.67}$$

where m is the usual discrete spectrum (2.26b) with constant relaxation time and $h(t,t')$ is Wagner's empirically determined 'damping function' which depends on the invariants of $\underline{C}_t^{-1}(t')$, and reduces the contributions to the stress from higher strains. Wagner finds that data is fitted best in elongation with a different damping function than in shear. If we write $\gamma = \log L(t)/L(t')$, the function

$$h(t,t') = 1/\{a \exp(2\gamma) + (1-a) \exp(m\gamma)\} \tag{5.68}$$

was used, with m = 0.3 and a = exp (-6).

Equal success in finding agreement between theory and experiment can be claimed by Acierno, La Mantia and Marrucci [A5,A7]. The merit of this approach is that the assumptions and the choice of the form of the material functions are made in a model which is related to a model of behaviour at a molecular level. This makes it more difficult to adjust the model if its predictions are not too good - which is not necessary on the evidence presented - but may just be disguising arbitrary choices in language unfamiliar to the solver of fluid dynamical problems.

Two topics which deserve some discussion are surface tension and the stability of stretching. The latter has only recently began to receive attention [A9] (see also [P3], as well as references cited in [A9]). Under some conditions the surface of a cylinder of molten polymer becomes very irregular while it is being stretched. We have discussed related problems elsewhere (sections 1.1.3, 1.2.3 and 6.4). The theoretical trick for the Newtonian liquid is to change independent variables to $\theta \equiv \log(L(t)/L_o)$ and $\zeta \equiv z/L(t)_o$. Then dependent variables are made dimensionless using the velocity $\dot{L}(t)$, the viscosity 3η and the specimen area $A(t)/A_o$ is used to define

$$D(\zeta,\theta) \equiv A(t)e^{\theta}/A_o. \tag{5.69}$$

This gives variables (D,U,T) whose values in uniform stretching are independent of time. Then a conventional stability analysis can be performed (as in section 6.4).

Surface tension may be expected to be relevant to instability, but this has not been investigated. What has been done by Laun and Münstedt [L5] is to calculate the effect surface tension (or rather interfacial tension) will have in a way similar to that outlined for spinning in section 3.3.1

(equations (3.2c), (3.6), (3.7)). Their deduction is based on calculation of the increase in interfacial energy due to the stretching, and leads to essentially the same equation. The measured stress is $F/\pi R^2$ for a circular specimen, and this consists of the rheological contribution due to the deformation of the material, $p_{11}-p_{22}$, and the contribution from surface tension σ/R, from (3.7) divided by πR^2,

$$F/\pi R^2 = p_{11} - p_{22} + \sigma/R \qquad (5.70)$$

The expression obtained by White and Ide [W13, equation (29)] appears to be in error because of the neglect of the contribution $2\pi R\sigma$ to the axial force at a cross-section due to the surface tension acting along the cylindrical surface of the specimen. They have (in our notation)

$$T_X = p_{11} - p_{22} - \sigma/R \qquad (3.5)$$

and T_X is correctly the stress at any point in the cross-section. However the total force requires to balance both $T_X\pi R^2$ and $2\pi R\sigma$, whence we obtain (5.70). This is in agreement with Laun and Münstedt [L5] who also have evidence from experiments at very low strain rates which confirm this extremely well. At low strain rates the polyethylene melt behaves as a Newtonian liquid, and the graph of stress against strain rate is linear from strain rates of $10^{-3}s^{-1}$ down to $10^{-4}s^{-1}$. Below this rate the stress exceeds the viscous stress by a measurable amount and tends to a limiting value of σ/R as the strain rate gets very small.

5.3.2 Biaxial stretching

This topic, like others we have dealt with briefly, is of enough importance to fill a chapter, but there has not been enough work done to make this necessary. We have discussed simple ideas in section 2.3, with steady flow results in section 2.2, and experimental methods in section 3.2. Two theoretical studies merit further attention here.

White [W12] discusses the kinematics of general (unequal) biaxial stretching of sheets of material and uses the Bogue-White single-integral constitutive equation in a simplified form which is equivalent to (2.36) if the rate of strain is constant. A noteworthy feature of the analysis is that in equal biaxial extension at constant strain rate the dominant term at large elongation rates arises from the term $\frac{1}{2}\varepsilon \underline{C}_t(t')$ in the integral (2.29). The stress extending the sheet takes the sign of $-\frac{1}{2}\varepsilon$ in this situation, so that we must have $\varepsilon \leq 0$ to get a positive (tensile) stress causing stretching. This is consistent with the current view that all reliable data on the second normal stress difference show it to be negative (see Ψ_2 in table 2.3 for the Ward-Jenkins model).

White draws attention to the fact that the large strain rate asymptotic form of the equations corresponds to the response of an elastic material (with the memory integral constitutive equation being replaced by a modulus multiplying the measure of deformation in (2.29)). This model has been used in calculations on the unequal biaxial stretching in processes for film manufacture. For application to a molten polymer in the conventional film blowing process [P12] it is obtained by assuming that the relaxation time is large in a Maxwell model (relaxation time multiplied by strain rate will give a dimensionless number - a Weissenberg number, and elastic effects predominate if this is much larger than one). In solid film stretching an empirically determined constitutive equation similar to these has been used [P6,G24], although some doubts have been expressed concerning the validity of the equation used in this case [P19]. While we are reviewing work of theoretical significance associated with film blowing we should mention the work of Wagner [W1] who has modelled the viscoelastic nature of the melt (rather than making viscous or elastic approximations) and Han and Park [H4], who use a purely viscous non-Newtonian model and determine "apparent extensional viscosities". This latter procedure is somewhat dangerous, as we have seen, since it ignores the elasticoviscous nature of the molten polymers, and the numbers obtained have to be treated with great caution.

White [W12] also considers biaxial extension with velocity of the specimen edges prescribed, and obtains results involving exponential integrals (as in uniaxial stretching [P16]). Finally, White considers the effect of inhomogeneities in the sheet, in an analysis based on that of Chang and Lodge [C7]. He deduces that even in a Newtonian material a small thin spot will not get thinner any more rapidly than the rest of the material. The analysis is clearly a much simplified treatment of an interesting problem which deserves further attention.

The second theoretical study referred to above is due to Acierno, La Mantia and Titomanlio [A6] who apply the constitutive equation (2.59) to the prediction of stress growth and recoil (recovery) which has been measured by Denson and his coworkers [D16,J12,J13] for polyisobutylene. The results of the comparisons of theory and experiment are favourable to the model, and it is particularly encouraging that the adjustable parameter, a, takes the same value for a different polyisobutylene [A8], and that time-dependent biaxial extension, uniaxial extension and shearing experiments are all considered with equal success. The parameter, a, is said to change only when there are important structural differences between polymers. In contrast to these calculations for idealised flow situations it is possible to calculate, for example, how a flat sheet is inflated into a bubble (as in biaxial extensional experiments). This has been done for a viscoelastic constitutive equation of an essentially solid-like nature by Wineman [W23] and there seems to be no obvious reason why this technique might not be extended to use liquid-like constitutive equations.

6 Theory of spinning

"You'll reply that reality hasn't the least obligation to be interesting. And I'll answer you that reality may avoid that obligation but that hypotheses may not."

Jorge Luis Borges.

The spinning problem we discussed in section 2.3.3 is much simplified, as we have already seen in discussing the analysis of experimental data (section 3.3.1). In the first section of this chapter we continue to ignore the real difficulties and discuss in detail the predictions of some constitutive equations. This is important since the analysis of the more difficult problems will almost certainly depend on the choice of constitutive equation and we should prefer to avoid devoting a great deal of effort to a model which has serious defects (as for example we find for the co-rotational model).

The second section offers some thoughts on the extremely important question of boundary conditions for spinning, while continuing to assume the simplified kinematics of steady uniaxial extension. The initial conditions clearly depend on the flow emerging from the die (orifice or capillary; spinnerette in commercial spinning). The transition from shear flow to elongational flow and the phenomenon of extrudate swell are clearly relevant here. The effect of the take-up is in some ways more important; the correct formulation of the problem as a two-point boundary-value problem alters its character markedly. A clear example of this is given by the answers to the question of the existence of solutions for the Newtonian liquid when inertia is significant (sections 6.1.6, 6.2.2). The initial value problem has no solutions beyond a certain distance, a fact which has been noted by several authors, but the correctly posed boundary-value problem always has a solution.

Some of the more subtle problems are discussed in the third section of this chapter, with the simplification that we restrict attention to Newtonian liquids. The problem of the variation of axial velocity across the jet cross-section and of the necessary presence of small shear stresses have not been fully explored. This does leave the earlier work on somewhat uncertain foundations, at least in the sense that we are not sure how slowly the jet radius must be changing for the simple kinematics to be a good approximation. Denn (unpublished) has given the matter some thought, viewing the approximation as an averaging process across the jet cross-section. Here we shall do no more than mention some relevant results and hope that the problems are clearly indicated. We do not believe (at present) that the shakiness of the foundations should lead to serious doubts concerning the analysis of spinning flows and the interpretation of experimental results.

In the final section on stability, we select two ideas from a very considerable amount of published work. The question of boundary conditions again arises and is extremely important, although we cannot do much in the present state of the art except outline the important issues. We conclude these introductory remarks by pointing out, in fairness to the reader, that some of the ideas and results presented here have not been published elsewhere and are occasionally speculative rather than thoroughly tested.

6.1 THE SIMPLIFIED INITIAL VALUE PROBLEM

For the moment we continue to ignore gravity, surface tension and inertia and to assume a flow which is uniform across a cross-section of the thread being spun. The basic problem is specified by (2.79) and (2.80), which give

$$FU(X)/Q = p_{11}(X) - p_{22}(X) \tag{6.1}$$

for known constant force F and flow rate Q. The extra-stress is given by a constitutive equation, and the rate of strain tensor we use (in this approximation) is (2.74), i.e. the principal rates of strain are $U'(X)$, $-\frac{1}{2}U'(X)$ and $-\frac{1}{2}U'(X)$.

6.1.1 Purely viscous constitutive equations

If we take a fluid with a power-law viscosity (2.39), giving

$$p_{11} - p_{22} = 3C(II_e)^n U'(X) \tag{6.2a}$$

with

$$II_e = (3/4)U'(X)^2 \tag{6.2b}$$

we obtain

$$U(X) = U(0)\{1 + KX\}^{1/m} \tag{6.3a}$$

where

$$m = 2n/(2n + 1) \tag{6.3b}$$

(the same exponent m as in Matovich and Pearson [M17, equation (47)] and

$$K = \left(\frac{4}{3}\right)^{m/2} \left(\frac{F}{3CQ}\right)^{1-m} m(U(0))^{-m} \tag{6.3c}$$

The Newtonian result may be recovered in the limit as $n \to 0$, and as $n \to \infty$ ($m \to 1$) the velocity profile becomes linear.

We note that the behaviour for $n > 0$ (materials which resist deformation more at higher rates of strain) is more like that observed for molten polymers and polymer solutions in elongation. Values in the range $-\frac{1}{2} < n < 0$, which is the range of power-law index that gives the commonly observed (shear thinning) behaviour of polymer melts and solutions in simple shear, are less satisfactory for elongation, and indeed we have $U \to \infty$ and $A \to 0$ as $X \to |K|^{-1}$. (Since $m < 0$ we have $K < 0$ and note also that since $1 - m > 0$, the distance over which solutions exist decreases as the applied force, F,

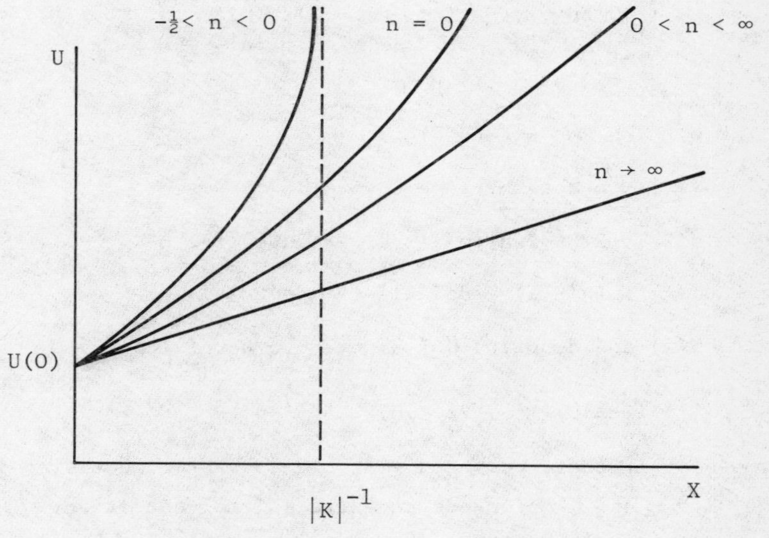

Figure 6.1 *Sketch of velocity profiles for power-law fluids (6.3).*

increases but increases as U(0), A(0) or C increase. The velocity profiles in Figure 6.1 have initial slope $U'(0) = (4/3)^{m/2}(FU(0)/3CQ)^{1-m}$.)

We shall not deal at further length with inelastic fluids, since it seems clear from all available evidence that elasticity plays a significant, if not a dominant role, in the spinning of polymeric liquids. Such an argument, based on consensus rather than exhaustive analysis, is clearly not conclusive, but is surely adequate reason for exercising a choice in studying models within practical limitations of time (and space for presentation). There is some discussion of this model by Hyun and Ballman [H21].

6.1.2 The three-parameter Maxwell model

The upper convected and corotational Maxwell models, each with just two parameters, have been discussed by Denn, Petrie and Avenas [D12] and by the author [P15] respectively. Both are contained in the three parameter model (2.21) which leads to

$$p_{11}(X) + \lambda\{U(X)p_{11}'(X) - 2aU'(X)p_{11}(X)\} = 2\eta U'(X) \qquad (6.4a)$$

$$p_{22}(X) + \lambda\{U(X)p_{22}'(X) + aU'(X)p_{22}(X)\} = -\eta U'(X) \qquad (6.4b)$$

and hence, with dimensionless variables

$$T(x) = p_{11}(X)Q/FU(0), \quad u(x) = U(X)/U(0), \quad x = X/L, \qquad (6.5a,b,c)$$

parameters

$$\alpha = \lambda U(0)/L, \ \mu = \eta Q/\lambda FU(0) \qquad (6.6a,b)$$

and the use of (6.1) we obtain

$$T(x) + \alpha\{u(x)T'(X) - 2au'(x)T(x)\} = 2\alpha\mu u'(x) \qquad (6.7a)$$

$$u(x) + \alpha u'(x)\{(1+a)u(x) - 3aT(x)\} = 3\alpha\mu u'(x) \qquad (6.7b)$$

By eliminating T(x) and denoting $\alpha u'(x)$ by $\dot{u}$ we have, for $a \neq 0$,

$$u\ddot{u} = \dot{u}(1 + (1+a)\dot{u})(1 - (2a-1)\dot{u}) - 3\mu\dot{u}^2/u \qquad (6.8)$$

We see immediately that, compared with the inelastic models, this model is of order one higher and hence needs an extra initial condition. This, as is perhaps best seen from (6.7a), is most easily thought of as an initial value for the extra-stress component T(x), which is a natural quantity to specify for a material exhibiting stress relaxation. Clearly, from (6.7b) it is alternatively possible to specify the initial value of $u'(x)$. Either of these extra conditions is required because the Maxwell model is a viscoelastic model (with a fading memory) and we need to specify something of the material's history prior to the spinning flow, which will be adequate to describe the influence, through the fading memory, of this prior history.

Now (6.8), or the equivalent system (6.7) does not have an explicit solution for $a \neq 0$ (unless $\alpha = 0$ or $\mu = 0$) and three investigations have been carried out, in addition to numerical solution for a limited number of cases. The small α (nearly Newtonian) and small μ (large applied force) solutions can be obtained, as an extension of [D12] (for $a = 1$) and give us respectively a singular perturbation problem and a regular perturbation problem. (In the Newtonian case, $\alpha = 0$, $\mu \to 0$ gives a singular problem.) Thirdly we can make some deductions concerning the qualitative nature of solutions for all α, μ and a, and from this we may be led to bounds on solutions and other semi-quantitative results of analysis.

We shall discuss the 'nearly Newtonian' approximation below (section 6.1.4) since this has features in common with the second-order fluid solutions. The large force approximation leads to some interesting, and unexpected, results which are exemplified by the co-rotational model. With $a = 0$ and $\mu = 0$ we obtain

$$T + \alpha u T' = 0 \tag{6.9a}$$

$$u + \alpha u u' = 0 \tag{6.9b}$$

so that we have $u' = -1/\alpha$, which we are disposed to reject as physically unreasonable. Some light is shed on this by looking at the full equations for the co-rotational model [P15]. We find

$$(3\mu - u)\,\alpha u' = u \tag{6.10}$$

which has a solution with $u'(0) > 0$ if $3\mu > u(0)$ and that solution will exist until $u(x)$ reaches the value 3μ. Since (6.10) has an implicit solution (satisfying $u(0) = 1$)

$$3\mu \log u - (u-1) = x/\alpha \tag{6.11}$$

we can write down the interval of existence

$$0 \leq x < x_c = \alpha\{3\mu \log 3\mu + 1 - 3\mu\} \tag{6.12}$$

We see that x_c increases with μ from a value of zero for $\mu = 1/3$ and $x_c \to \infty$

as $\mu \to \infty$. For $\mu < 1/3$ we have $x_c > 0$ again, but solutions now have $u' < 0$. Thus as the applied force is increased from zero (μ infinite) to $3\eta Q/\lambda U(0)$ the solution breaks down at a finite value of u in a distance x_c which decreases to zero, and if we reject solutions with $u' < 0$ (the filament thickens in response to an applied tension) we have a maximum allowed force which corresponds to an initial stress $F/A(0) = 3\eta/\lambda \equiv 3G$, i.e. equal to the tensile modulus of the material.

For the rest of the class of models we could find the $\mu = 0$ solution by integrating

$$\left\{\frac{(1+a)}{1+(1+a)\dot{u}} + \frac{(2a-1)}{1-(2a-1)\dot{u}}\right\} \frac{d\dot{u}}{du} = \frac{3a}{u} \tag{6.13}$$

which is obtained by rearranging (6.8). Provided that the denominators are one-signed this gives

$$\dot{u} = (Ku^{3a}-1)/\{1+a+(2a-1)Ku^{3a}\} \tag{6.14}$$

which is integrated for $a = 1$ [D12]. K is determined by the initial value of $\dot{u}$ or equivalently of T. It is helpful here to consider the $(u,\dot{u})$ phase-plane for the equation (Figure 6.3, below), and to treat separately the ranges $\frac{1}{2} < a \leq 1$, $0 < a < \frac{1}{2}$, $-1 < a < 0$, with $a = \frac{1}{2}$, $a = -1$ (and $a = 0$, which we have already discussed) as special cases.

Clearly $\ddot{u}$ is zero when $\dot{u} = 0$, $-1/(1+a)$ and $1/(2a-1)$. Hence for $\frac{1}{2} < a \leq 1$ there are two regions in the quadrant $u > 0$, $\dot{u} > 0$ of the phase plane, namely $0 < \dot{u} < 1/(2a-1)$ where $\ddot{u} > 0$ and $\dot{u} > 1/(2a-1)$ where $\ddot{u} < 0$. Hence as x increases $\dot{u} \to 1/(2a-1)$ and $\ddot{u} \to 0$, so we eventually get a linear velocity profile $u = k + x/(2a-1)$ and steady stretching at a strain rate $U'(X) = 1/(2a-1)\lambda$. Numerical experience for $a = 1$ [D12] suggests that this limit is attained quite rapidly unless extremely large values of $T(0)$ are prescribed.

When $-1 < a < \frac{1}{2}$, $\ddot{u} > 0$ always for $\dot{u} > 0$ and $u > 0$, so $\dot{u}$ increases without bound, and we are led to enquire about existence of solutions, noting the finite interval of existence we have for $a = 0$. From (6.14) we may show that for $0 < a < \frac{1}{2}$ and $\dot{u}(0) > 0$ we have $1 < K < (1+a)/(1-2a)$. We see then that as u increases from its initial value of one, u^{3a} will increase towards the value

$$(1+a)/\{(1-2a)K\} \tag{6.15}$$

and will reach it in a distance which we could bound above by bounding $\dot{u}$ below. This then means that we have a finite interval of existence for any $\dot{u}(0) > \delta$, since as u^{3a} approaches the value (6.15), $\dot{u}(x) \to \infty$. A similar argument holds for $-1 < a < 0$, when $0 < (1+a)/(1-2a) < K < 1$, and as u increases u^{3a} decreases towards the same critical value (6.15). One disturbing feature of the situation for $-1 < a < 0$ is that T(0), obtained from (6.7b) as $\{1 + (1+a)\dot{u}(0)\}/\{3a\dot{u}(0)\}$ is negative for positive $\dot{u}(0)$. This appears to be physically unacceptable, and is strong evidence against the applicability of the model with $a < 0$.

The argument, at least as far as qualitative behaviour of solutions is concerned, can be extended to non-zero μ. Basically we can divide the phase-plane into regions where $\ddot{u} > 0$ and $\ddot{u} < 0$. For $\frac{1}{2} < a \leq 1$ there is a curve $\ddot{u} = 0$ in the first quadrant given by the positive root $\dot{u}_m(u)$ of

$$1 + (2-a-3\mu/u)\dot{u} - (1+a)(2a-1)\dot{u}^2 = 0 \qquad (6.16)$$

(obtained from (6.8)), and as $u \to \infty$, $\dot{u}_m \to 1/(2a-1)$ from below. Then any physically relevant solution curve (i.e. with $u(0) = 1$, $\dot{u}(0) > 0$) must follow a trajectory for which u(t) always increases and $\dot{u}(t)$ decreases until (and if) it reaches a value $\dot{u}_m(u) < 1/(2a-1)$ for some u and then will increase, but will be bounded above by $\dot{u}_m(u)$.

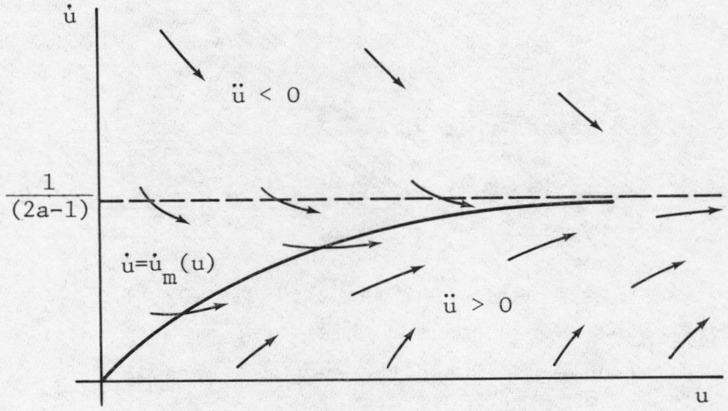

Figure 6.2 Phase plane for the Maxwell model (6.8) for ½ < a ≤ 1.

For $a < \frac{1}{2}$ and u large enough, $\ddot{u} > 0$ for all $\dot{u} > 0$ and hence u and $\dot{u}$ increase monotonically. If $\dot{u}$ is large (and u is not small), we could approximate (6.8) by

$$u\ddot{u} = (1+a)(1-2a)\dot{u}^3 \tag{6.17}$$

with the solution

$$\dot{u} = \frac{1}{K-(1+a)(1-2a)\log u} \tag{6.18a}$$

$$u\{K + (1+a)(1-2a)(1-\log u)\} = x/\alpha \tag{6.18b}$$

for which $\dot{u} \to \infty$ at finite values of x and u. This suggests, at least, that for $a < \frac{1}{2}$, (6.8) has solutions only over a limited interval in x.

Apart from a = 0, no special cases lead to obvious simplifications though $a = \frac{1}{2}$, as the critical value separating the two types of behaviour is perhaps worth a little further discussion. This (bifurcation) value has the equation

$$u\ddot{u} = \dot{u}\{1 + 3\dot{u}(\tfrac{1}{2}-\mu/u)\} \tag{6.19}$$

so $\ddot{u} > 0$ for $u > 6\dot{u}\mu/(2+3\dot{u})$.

Hence we may deduce that u increases monotonically from the initial value $u(0) = 1$ for $\dot{u}(0) > 0$, and eventually $\mu/u \ll \frac{1}{2}$, so we write

$$u\ddot{u} \sim \dot{u}(1+3\dot{u}/2). \tag{6.20}$$

This has the solution (if it holds for $x \geq 0$)

$$\frac{x}{3\alpha} = \frac{1}{k^2}\int_k^{k\sqrt{u}} \frac{t\,dt}{t^3-1} \tag{6.21}$$

where $k^3 = 1 + 3\dot{u}(0)/2$, and we see that $u \to \infty$ as

$$x \to \frac{3\alpha}{k^2}\int_k^{\infty} \frac{t\,dt}{t^3-1} = \frac{3\alpha}{k^2\sqrt{3}}\left(\frac{\pi}{2} - \tan^{-1}\left(\frac{2k+1}{\sqrt{3}}\right)\right) + \frac{\alpha}{6k^2}\log\left(\frac{k^3-1}{(k-1)^3}\right) \tag{6.22}$$

Since in fact $\ddot{u}$ is smaller, the solution does not break down as quickly as this, and (6.22) gives us a lower bound for this interval of existence. We note that here (unlike the case a = 0 and the supposed behaviour for $a < \frac{1}{2}$) $\dot{u}$ and u both tend to infinity at the upper limit of the interval of existence.

The main practical conclusion from this exercise is that the simple Maxwell model is to be treated with extreme caution if $0 \leq a \leq \frac{1}{2}$ (which includes the co-rotational model), and avoided altogether for $a < 0$.

6.1.3 The four-parameter Jeffreys model

The mathematical problems encountered in the use of simple Maxwell models are of the type which may be overcome by the introduction of a higher derivative into the governing equations. Hence we add a retardation time multiplying a time derivative of the rate of strain to give constitutive equation (2.23). Such models have been studied less extensively than the corresponding (simpler) Maxwell models [P18]. Equations (6.4) are modified by the addition of terms

$$2\eta\Lambda\ \{U(X)U''(X) - 2a(U'(X))^2\} \tag{6.23a}$$

and

$$-\eta\Lambda\ \{U(X)U''(X) + a(U'(X))^2\} \tag{6.23b}$$

to their respective right-hand sides. Using the dimensionless ratio $\beta = \Lambda/\lambda$ we have, corresponding to (6.7a)

$$T + \alpha\{uT' - 2au'T\} = 2\alpha\mu[u' + \alpha\beta\ \{uu'' - 2au'^2\}], \tag{6.24a}$$

$$u + \alpha u'\ \{(1 + a)\ u - 3aT\} = 3\alpha\mu[u' + \alpha\beta\ \{uu'' - au'^2\}]. \tag{6.24b}$$

It is simplest to discuss special cases, since a full study of this system, or the equivalent third-order equation (which corresponds to (6.8); $\dot{u} = \alpha u'$)

$$\begin{aligned} 3\beta\mu u^2\dot{u}\,\dddot{u} = {} & -u^2\ddot{u} + u\dot{u}\ \{1+(1+a)\ \dot{u}\}\{1 - (2a-1)\ \dot{u}\} - 3\mu\dot{u}^2 \\ & + 3\beta\mu\ \{a\dot{u}^3 - u\dot{u}\ddot{u} + u^2\ddot{u}^2 + 2a^2\dot{u}^4 - (1-a)\ u\dot{u}^2\ddot{u}\} \end{aligned} \tag{6.25}$$

is a somewhat daunting prospect. We consider first the co-rotational model, where $a = 0$ reduces (6.24b) to

$$3\beta\mu u\ddot{u} + (3\mu - u)\dot{u} - u = 0. \tag{6.26}$$

For $u > 1$ there is always a neighbourhood of $\dot{u} = 0$ where $\ddot{u} > 0$, so that if $\dot{u}(0) \geq 0$ then $\dot{u}(x) > 0$ for $x > 0$, for u increases monotonically with x. For $u < 3\mu$, if $0 \leq \dot{u} < u/(3\mu-u)$ we have $\ddot{u} > 0$ while if $\dot{u} > u/(3\mu-u)$ we have $\ddot{u} < 0$, so that $\dot{u}$ may decrease initially, but whatever its initial value it will eventually increase monotonically. Thus when $u \gg 3\mu$ and $\dot{u} \gg 1$, as will happen eventually

$$\ddot{u} \sim \dot{u}/3\beta\mu$$

and hence

$$\dot{u} \sim u/3\beta\mu$$

$$u \sim k \exp (x/3\beta\mu\alpha) , \qquad k \text{ is a constant} \tag{6.27}$$

or $$U \sim K \exp\{(FX/3\eta Q)(\lambda/\Lambda)\} , \qquad K = kU(0),$$

which has the same behaviour as a Newtonian liquid with viscosity $\eta\Lambda/\lambda$, consistent with the idea of an "upper Newtonian" region of behaviour at high strain rate.

We find that, as expected, there are now no existence difficulties; (6.26) has solutions for all x. To show this note that the coefficient of $\ddot{u}$ never vanishes, and u and $\dot{u}$ may be bounded for bounded x.

We bound $\dot{u}$ by writing (6.26) in the form

$$\ddot{u} = \frac{1}{3\beta\mu} + \frac{\dot{u}}{3\beta\mu} - \frac{\dot{u}}{\beta u}$$

$$\leq \frac{\dot{u}+1}{3\beta\mu}$$

if $\dot{u} > 0$, $u > 0$ and $\beta > 0$. Then integration gives

$$\dot{u}(x) \leq Ke^{x/3\alpha\beta\mu} - \alpha$$

where $K = \dot{u}(0) + \alpha > 0$, and hence for any bounded x, $\dot{u}(x)$ is bounded. A further integration gives a similar bound for u.

The asymptotic result (6.27) suggests that we try something similar for the more general model and hence seek solutions of the form

$$u \sim Ae^{kx/3\beta\mu\alpha}$$

for large A (or large x). We find that the left-hand side of (6.25) is $k^4u^4/(3\beta\mu)^3$ and the right-hand side is given by

$$\frac{k^4u^4}{(3\beta\mu)^3}\left(2a^2 + a - \frac{(1+a)(2a-1)}{k}\right) + \frac{k^3u^3}{(3\beta\mu)^2}\left(a - 1 + \frac{1-a}{k}\right) + \frac{k^2u^2}{3\beta\mu}\left(\frac{1}{k} - \frac{1}{\beta}\right) \tag{6.28}$$

and if $ku/3\beta\mu >> 1$ we require, for equality of the leading terms,

$$2a^2 + a - (1+a)(2a-1)/k = 1$$

or

$$k = 1. \tag{6.29}$$

Thus we find the same "upper Newtonian viscosity", $\mu\Lambda/\lambda$, as is obtained in steady shear flow with the same model.

It must be emphasised that at this stage there is no proof that the large u solutions are ever attained, except in the case of the co-rotational model (which is rather special). However this behaviour does at least rule out breakdown of solutions as encountered with Maxwell models for $a < \frac{1}{2}$. More careful argument concerning existence of solutions of (6.25) tells us that there is at least one solution through every point in the domain

$$-\infty < \ddot{u} < \infty,\ 0 < \dot{u} < \infty,\ 0 < u < \infty,\ -\infty < x < \infty,$$

and that a solution may be continued up to the boundary of this domain (Peano existence theorem and corollary, see for example [H1] pp. 14,17). Hence a solution which starts with $u > 0$, $\dot{u} > 0$, $\ddot{u}$ bounded at $x = 0$ will exist for all x unless u or $\dot{u}$ can approach zero, or tend to infinity. We thus need to investigate solutions for small u and for small $\dot{u}$, and need to bound solutions above as we did for the co-rotational model.

In addition to the upper Newtonian limiting solution (6.27), there is another approximation solution that we can obtain. If u is large but $\dot{u}$ is not large the values $\dot{u} = 1/(2a-1)$, $\ddot{u} = \dddot{u} = 0$ satisfy (6.25) with error $\ll u$ for large u. This corresponds to the limiting strain rate for the Maxwell model for $\frac{1}{2} < a \leq 1$. The question remains open as to how $\dot{u}$ behaves as u gets large; at present we only have a limited amount of computational evidence to suggest that $\dot{u}$ will tend to $1/(2a-1)$ rather than $u/3\beta\mu$. It may also be noted that there is, perhaps, a contradiction between the attainment of the "upper Newtonian" regime and the existence of a limiting strain rate for steady flow, but this is at present a speculation.

6.1.4 Retarded motion expansions: "nearly Newtonian" fluids

The second order fluid (3.27) gives stresses

$$p_{11}(X) = 2\eta U'(X) - 2\theta(U(X)U''(X) + 2U'(X)^2) + 4\chi U'(X)^2 \quad (6.30a)$$

$$p_{22}(X) = -\eta U'(X) + \theta(U(X)U''(X) - U'(X)^2) + \chi U'(X)^2 \quad (6.30b)$$

and hence, using (6.1), we obtain

$$FU(X)/Q = 3\eta U''(X) - 3\theta U(X)U''(X) + 3(\chi-\theta)U'(X)^2 . \quad (6.31)$$

This may also be obtained from equation (53) of Matovich and Pearson [M17] by integrating (after dividing by v, correcting the first v"' to v" and dropping the inertia, gravity and surface tension terms), and then F/Q appears as a constant of integration.

An alternative derivation of this, which shows that the same approximation is obtained from the Maxwell model (2.21) when $\lambda U'(X)$ is small, is to approximate (6.4a) by $p_{11}(X) = 2\eta U'(X) + O(\lambda U'(X))$ and then, as a second approximation, to use this Newtonian first approximation in the terms multiplied by λ in (6.4a) to get

$$p_{11}(X) = 2\eta U'(X) - 2\eta\lambda\{U(X)U''(X) - 2aU'(X)^2\} + O(\lambda^2 U'(X)^2)$$

and a similar expression for $p_{22}(X)$. These may be identified with (6.30) if we set $\theta = \eta\lambda$ and $\chi - \theta = a\eta\lambda$. The dimensionless form of (6.31) is obtained by defining as before (6.5,6.6) $u = U/U(0)$, $x = X/L$, $\alpha = \theta U(0)/\eta L$, $a = (\chi-\theta)/\theta$ and $\mu = \eta^2 Q/\theta\, FU(0)$. We have then

$$3\mu\alpha u'(x) - u(x) = 3\mu\alpha^2\{u(x)u''(x) - au'(x)^2\} \tag{6.32}$$

which can be obtained from [M17 , equation (55)] by dividing by ψ^2 and integrating, with constant of integration $1/3\mu\alpha$, and ψ, Δ, ξ corresponding to our u, α, a.

We emphasise the comparison because the subsequent discussion by Matovich and Pearson contains an important alternative procedure for developing solutions of (6.32) for small α. Denn, Petrie and Avenas [D12] were not able to show the correspondence between their work and that of Matovich and Pearson. There are some unfortunate misprints in the 1969 paper which obscure the point at issue, and of course it is only in the limiting case of small α that we expect the same result for second order and Maxwell fluids, but in this limit the equations of both papers are exactly equivalent.

The singular perturbation problem for small α is tackled from the system (6.7), noting that the nearly Newtonian limit $\lambda \to 0$ corresponds to $\alpha \to 0$ and $\mu \to \infty$. We therefore introduce the O(1) parameter

$$\varepsilon = \alpha\mu = \eta Q/LF \tag{6.33}$$

Then we have

$$3\varepsilon u' - u = \alpha((1+a)uu' - 3au'T) \tag{6.34}$$

and eliminating the $\varepsilon u'$ term between the equations,

$$3T - 2u = \alpha(2(1+a)uu' - 3uT). \tag{6.35}$$

Hence, with $u = u_o + \alpha u_1 + \ldots$, $T = T_o + \alpha T_1 + \ldots$, and letting $\alpha \to 0$ with x fixed, we have

$$u_o = U_o e^{x/3\varepsilon}, \tag{6.36a}$$

$$T_o = 2U_o e^{x/3\varepsilon}/3, \tag{6.36b}$$

the Newtonian solution as our zero-order approximation.

The first-order correction is

$$u_1 = U_1 e^{x/3\varepsilon} + (1-a)U_o^2 e^{2x/3\varepsilon}/3\varepsilon \tag{6.37a}$$

$$T_1 = \tfrac{2}{3}U_1 e^{x/3\varepsilon} + 2U_o^2 e^{2x/3\varepsilon}/3\varepsilon \tag{6.37b}$$

and the second-order terms are

$$u_2 = U_2 e^{x/3\varepsilon} + 2(1-a)U_o U_1 e^{2x/3\varepsilon}/3\varepsilon + (3-6a+a^2)U_o^3 e^{x/\varepsilon}/18\varepsilon^2 \tag{6.38a}$$

$$T_2 = 2U_2 e^{x/3\varepsilon}/3 + 4U_o U_1 e^{2x/3\varepsilon}/9\varepsilon + (3-4a-5a^2)U_o^3 e^{x/\varepsilon}/27\varepsilon^2. \tag{6.38b}$$

The constants U_o, U_1, U_2 have to be determined by matching at $x = 0$ with an inner expansion which allows the conditions $u(0) = 1$ and $T(0) = \tau$ both to be satisfied. The above outer expansion satisfies $u(0) = 1$ if

$$U_o = 1, \; U_1 = -(1-a)/3\varepsilon, \; U_2 = (1-2a+3a^2)/18\varepsilon^2,$$

and then

$$T(0) = (2/3)[1 + \alpha a/3\varepsilon - \alpha^2 a(1+a)/9\varepsilon^2 + O(\alpha^3)].$$

We note that for $a = 0$ (the co-rotational model) the velocity field may be determined from (6.7b) independently of the stress field, which satisfies (6.7), conveniently re-written

$$\alpha u(3T-2u)' + (3T-2u) = 0, \tag{6.39}$$

and hence clearly $T(x) = 2u(x)/3$ if $T(0) = 2u(0)/3$, and otherwise $3T-2u \to 0$ as $x \to \infty$. (This provides a convenient partial check on our perturbation solution for arbitrary a.)

For the inner solution we use a stretched coordinate $\xi = x/\alpha$, and let $a \to 0$ with ξ fixed, so that the region $\xi = O(1)$ has $x = O(\alpha)$ giving us, in effect, a boundary layer near x = 0. The equations become

$$3\varepsilon\dot{u} = \alpha[u+\dot{u}\{(1+a)u-3aT\}] \qquad (6.40a)$$

$$3(u\dot{T}+T) = 2u[1+(1+a)\dot{u}] \qquad (6.40b)$$

where $(\dot{\ }) = d(\ )/d\xi = \alpha(\ )'$. Then we write inner expansions

$$u(\xi) = v_o + \alpha v_1 + \alpha^2 v_2 + \ldots \text{ and } T(\xi) = S_o + \alpha S_1 + \alpha^2 S_2 + \ldots,$$

from which we find, using the boundary condition u(0) = 1, T(0) = τ,

$$v_o = 1 \qquad (6.41a)$$

$$S_o = 2/3 + (\tau-2/3)e^{-\xi} \qquad (6.41b)$$

$$v_1 = \xi/3\varepsilon \qquad (6.42a)$$

$$S_1 = \tfrac{2}{3}\{\xi/3\varepsilon+a(1-e^{-\xi})/3\varepsilon\} + (\tau-2/3)\xi^2 e^{-\xi}/6\varepsilon \qquad (6.42b)$$

$$v_2 = \xi^2/18\varepsilon^2 + (1-a)\xi/9\varepsilon^2 + (\tau-2/3)a(e^{-\xi}-1)/3\varepsilon^2 \qquad (6.43)$$

and S_2 is given by an expression occupying several lines, which we omit. In order to evaluate the constants in the outer expansions we re-write the outer expansion in terms of the inner variable and expand in powers of α. We find that the series agree to order α^2 if $U_o = 1$, $U_1 = -(1-a)/3\varepsilon$ and $U_2 = 2(1-a)^2/9\varepsilon^2 - (3-6a+a^2)/18\varepsilon^2 - (\tau-2/3)a/3\varepsilon^2$. Hence we may write down a composite expansion, which will be uniformly valid, by adding to the outer expansion the terms in the inner expansion which do not appear in the outer expansion, namely the $e^{-\xi}$ or $e^{-x/\alpha}$ terms. This gives us finally

$$u = e^{x/3\varepsilon} + \alpha(e^{2x/3\varepsilon}-e^{x/3\varepsilon})(1-a)/3\varepsilon + \alpha^2(e^{x/\varepsilon}-e^{x/3\varepsilon})(3-6a+a^2)/18\varepsilon^2$$
$$+ \alpha^2(e^{2x/3\varepsilon}-e^{x/3\varepsilon})(-2(1-a)^2/9\varepsilon^2) + \alpha^2(e^{-x/\alpha}-e^{x/3\varepsilon})(\tau-2/3)a/3\varepsilon^2 + O(\alpha^3) \qquad (6.44a)$$

$$T = \tfrac{2}{3}e^{x/3\varepsilon} + (\tau-2/3)e^{-x/\alpha} + \alpha 2(e^{2x/3\varepsilon}-e^{x/3\varepsilon})/9\varepsilon \qquad (6.44b)$$
$$+ \alpha 2a(e^{x/3\varepsilon}-e^{-x/\alpha})/9\varepsilon + (\tau-2/3)x^2 e^{-x/\alpha}/6\alpha\varepsilon + O(\alpha^2)$$

One obvious practical point should be made at this stage. The time-consuming development of these long expressions and the use of the matching process is required to fit the additional stress (or velocity gradient) initial condition. However, in practice this condition is not known and the details of the boundary layer between our arbitrary origin and the outer region are probably quite insignificant beside the details of the matching (in the die swell region) between the shear flow in the die and the elongational flow we have been analysing. Perhaps the matching process will be adaptable to the more realistic analysis of the whole flow once we have a better understanding of the flow in the die swell region, and for this reason it has some merit. It does also confirm our statements about the structure of the problem and shows how the initial stress can influence the flow.

It does seem, therefore, that the alternative development due to Matovich and Pearson is worth more serious attention than it has received hitherto. In our notation the idea is to consider α as a slowly varying function of x. Then seeking a solution of the form

$$u = C_1(x)e^{C_2(x)x} \tag{6.45}$$

where C_1 and C_2 are slowly varying functions of x leads to the equation (dropping terms in $\alpha C_1'(x)$ and $\alpha C_2'(x)$)

$$uu'' - u^2 + \alpha(a-1)u'^3 = 0 \tag{6.46}$$

whose solutions may be expressed implicitly by

$$K \log u - \alpha(1-a)u = x$$

where K depends on $u'(0)$. Further investigations of the approximation technique which, according to Pearson (unpublished) is equivalent to Lighthill's technique of strained coordinates, are clearly required. It may well be more fruitful in conjunction with the Maxwell or Jeffreys models.

However it should be noted that there are some problems associated with (6.46) and deductions made from it. Matovich and Pearson discuss the influence of a (or $(\chi-\theta)/\theta$) on spinnability, using the qualitative criterion that a material is spinnable if $u'(x) \to 0$ as $x \to \infty$ with $u(x)$ bounded (i.e. that an infinitely long thread of final diameter bounded away from zero is possible). Numerical results show that $a > 1$ aids spinnability (reduces $u'(x)$ at any x) and $a < 1$ does the reverse. Unfortunately we are disposed to reject $a > 1$ on other grounds (such as a positive second normal stress difference and viscosity increasing with shear rate cf. Table 2.2). The conclusion is consistent with that for the Maxwell model for which $u'(x) \to \{\alpha(2a-1)\}^{-1}$ as $x \to \infty$, and this limit decreases as a increases from $\frac{1}{2}$.

However even for $a > 1$ we do not have spinnability in the sense of Matovich and Pearson while their results appear to indicate that they do obtain solutions with $u'(x) \to 0$.

The trouble here may lie in the fact that solutions of (6.46) have a limited interval of existence which may be associated with the approximations used in deriving it. However there are similar problems with the full equation for the second-order fluid (6.32) with α not necessarily small. With $\dot{u} = \alpha u'(x)$ as before, we rewrite (6.32) in the form

$$u\ddot{u} = \dot{u} + a\dot{u}^2 - u/3\eta \tag{6.48}$$

The phase plane (Figure 6.3) has trajectories for which, if $a > 0$, u at first increases then decreases to zero (with $\dot{u} \to -1/a$). This is necessarily true if initially $u \geq 3\eta(\dot{u} + a\dot{u}^2)$, and it is not clear whether trajectories with smaller initial values of u (or larger initial $\dot{u}$) avoid the same fate. Thus some solutions (and perhaps all) do have a limited interval of existence and we cannot expect arguments based on large x behaviour to be very helpful. Once again, since the expected usefulness of the second-order fluid is for small α, we do not pursue this at length.

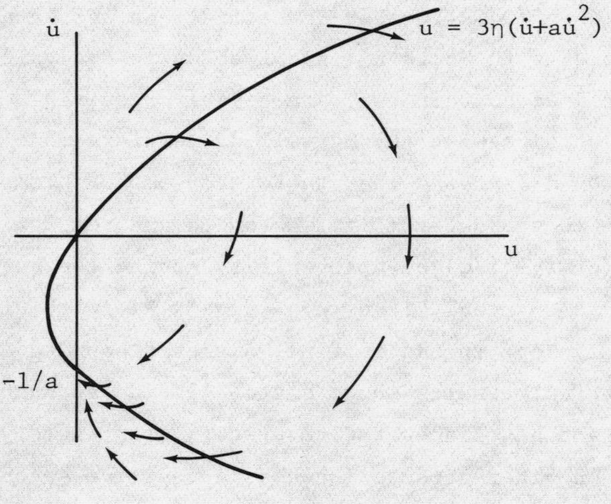

Figure 6.3 Phase plane for the second-order fluid (6.48) for $a > 0$.

We shall not discuss here the use of a third-order fluid model (see [M16]), since this has not seemed to promise much which is new and either mathematically interesting or practically relevant.

6.1.5 Rate equation models

There are many more complex models than those we have discussed so far in this chapter, but it becomes increasingly difficult to say anything general about solutions for such models, and we shall briefly review some published work. Denn and Marrucci [D13] have extended work on the upper convected Maxwell model to take account of a spectrum of relaxation times (cf. (2.22)) and Fisher and Denn [F9] have considered parameters which vary with the second invariant of the rate of strain tensor, using the White-Metzner model. In addition Denn [D12] formulated the problem for an extension of the upper convected Maxwell model, and here a phase plane analysis is informative.

The constitutive equation with a discrete relaxation spectrum is, in effect,

$$p_{11}(X) = \sum_{k=1}^{M} \tau_k(X) \tag{6.49a}$$

and

$$p_{22}(X) = \sum_{k=1}^{M} \pi_k(X) \tag{6.49b}$$

where τ_k and π_k are components of partial stress tensors $\underline{p}_k$ and here these satisfy

$$\tau_k(X) + \lambda_k\{U(X)\tau_k'(X) - 2U'(X)\tau_k(X)\} = 2\lambda_k G_k U'(X), \tag{6.50a}$$

$$\pi_k(X) + \lambda_k\{U(X)\pi_k'(X) + U'(X)\pi_k(X)\} = -\lambda_k G_k U'(X), \tag{6.50b}$$

and

$$\sum_{k=1}^{M} \{\tau_k(X) - \pi_k(X)\} = FU(X)/Q. \tag{6.51}$$

Note that we have now 2M first-order differential equations in the 2M unknowns τ_k and π_k (thinking of using the algebraic equation (6.51) to eliminate U(X) from (6.50). We therefore need 2M initial conditions, of which U(0) is the only kinematic one. The use of a spectrum to describe more accurately the elastic 'memory' of the material leads to the need for more initial conditions or, in other words, for a more detailed specification of the influence of the flow upstream of $X = 0$, i.e. of the flow which the fluid 'remembers' - perhaps this is natural for a fluid with a 'better' memory.

Numerical results [D13] show that the response for a spectrum typical of a low-density polyethylene melt is qualitatively similar to that of the model with a single relaxation time, both in the general shape of an (α,ε) graph and in the tendency towards a linear velocity profile and to a maximum value of α (for a given draw ratio; see below, Section 6.2) as ε tends to zero (the large force approximation). Here ε and α are based on mean values.

$$\overline{\eta} = \sum_{k=1}^{M} \lambda_k G_k, \quad \overline{\lambda} = \sum_{k=1}^{M} \lambda_k^2 G_k/\overline{\eta} \quad \text{and} \quad \overline{G} = \overline{\eta}/\overline{\lambda} \qquad (6.52a,b,c)$$

If we assume that a linear velocity profile, $U = BX + C$, is possible, in the limit of large force, when we neglect the terms on the right-hand sides, and substitute this into (6.50) we obtain 2M uncoupled equations with solutions

$$\tau_k(X) = \tau_k(0)\ (1+BX/C)^{(2\lambda_k B-1)/\lambda_k B} \qquad (6.53a)$$

$$\pi_k(X) = \pi_k(0)\ (1+BX/C)^{-(\lambda_k B+1)/\lambda_k B} \qquad (6.53b)$$

Then (6.51) will be satisfied for large X if the largest exponent in these expressions is equal to one, i.e. if

$$(2\lambda_m B-1)/\lambda_m B = 1; \quad \lambda_k < \lambda_m \quad \text{for } k \neq m \qquad (6.54a)$$

and also

$$\tau_m(0)/C = F/Q \qquad (6.54b)$$

The other terms in this sum in (6.51) will grow more slowly and hence may eventually be neglected. Of course with suitable initial conditions the linear velocity profile will satisfy (6.51) for all X; we require $\pi_k(0) = 0$ for all k and $\tau_k(0)$ for $k \neq m$. The rate of strain B is thus for large force, from (6.54a),

$$B = 1/\lambda_m \qquad (6.55)$$

for λ_m the largest eigenvalue is the spectrum. In practice it seems unlikely that the initial conditions will be appropriate for this exact solution, but we may still expect the solution to tend to this as X gets large, and hence predict a limiting strain rate which depends on the largest relaxation time, just as is the case for stretching at constant rate of strain, [P16]. This behaviour, we recall, is exactly the same as in the case of one relaxation time; the linear profile is only obtained as a limit for large X unless the initial conditions are exactly right.

Hence, in the dimensionless formulation [D13] we have a limiting strain rate

$$u'(x) \to \bar{\lambda}/\alpha\lambda_m \tag{6.56}$$

and if we suppose that $u'(x) < \bar{\lambda}/\alpha\lambda$ for $0 \leq x \leq 1$ we clearly have

$$\alpha(D_R - 1) < \bar{\lambda}/\lambda_m \leq 1 \tag{6.57}$$

with equality only for the case of a single relaxation time. The numerical results of Denn and Marrucci are in qualitative agreement with these predictions though not as close numerically as might be hoped. For their spectrum I and for M = 2 the limiting value of $\alpha(D_R-1)$ as $\varepsilon \to 0$ is between 0.73 and 0.74 and $\lambda/\lambda_m = 0.743$. However for M = 6 and spectrum II we have $\alpha(D_R-1) \to 0.66$ and $\bar{\lambda}/\lambda_m = 0.557$ and with spectrum IV, $\alpha(D_R-1) \to 0.49$ and $\bar{\lambda}/\lambda_m = 0.402$. The values of $\bar{\lambda}/\lambda_m$ appear to be much slower in tending to a limit (if indeed they ever do) than the values of $\alpha(D_R-1)$ as M increases. This disagreement may be attributable to choice of initial conditions, and clearly requires eluicidation if the qualitative analysis is to stand. Of course there are gaps in the analysis, since merely demonstrating the existence of a possible limiting solution does not prove that solutions tend to this limit or are bounded by it.

The analysis for the White-Metzner model (2.58) [F9] develops the theory to deal with materials with strain-rate dependent viscosities. The constant viscosity η is replaced by a function of the second invariant of the rate of strain tensor, and the function is chosen to give the popular power-law dependence of viscosity so that here

$$\eta(II_e) = C(\sqrt{3}\,U'(X))^{n-1} \tag{6.58}$$

The relaxation time is also allowed to vary, and is taken to be $\mu(II_e)/G$ with a constant modulus, G. The dimensionless group α is defined using this relaxation time with $U'(X) = U(0)/L$, as

$$\alpha = \frac{C}{G}\left(\frac{\sqrt{3}\ U(0)}{L}\right)^{n-1} \frac{U(0)}{L} \tag{6.59}$$

and $\varepsilon = \alpha\mu$ with μ given by (6.6b)

Then instead of (6.8) we have

$$\alpha n u u'' = u'^{2-n}(1+2\alpha u'^n)(1-\alpha u'^n) - 3\varepsilon u'^2/u \tag{6.60}$$

and we see that for $\varepsilon \to 0$ solutions will all have $\alpha u'^n \to 1$, and hence the limiting strain rate is $\alpha^{-1/n}$. Thus even if $n < 1$ (shear-thinning behaviour) the model predicts extension-thickening behaviour for large values of applied force.

It is an interesting mathematical exercise, but of limited rheological importance to look at solutions of the full equations considered by Denn [D12] which correspond to the Oldroyd model (2.25) with $\mu_1 = \lambda_1 = \lambda$, $\nu_1 = \nu_2 = \mu_o = \lambda_2 = 0$ and $\mu_2 = 2\lambda\nu$. This leads to

$$\begin{aligned}\alpha(u+6\alpha\varepsilon\nu u'^2)u'' &= u'(1+2\alpha u')(1-\alpha u') - 3\varepsilon u'^2/u \\ &\quad - 6\alpha\varepsilon\nu u'^3/u - 12\alpha^2\varepsilon\nu u'^4 u.\end{aligned} \tag{6.61}$$

We expect $\nu < 0$, since this corresponds to a negative second normal stress difference and to shear-thinning behaviour. Then writing $\alpha u' = \dot{u}$ as before and $k = -6\varepsilon\nu/\alpha$ we have

$$(u-k\dot{u}^2)\ddot{u} = u(1+2\dot{u})(1-\dot{u}) + k\dot{u}^2\ (1/2\nu+\dot{u}+2\dot{u}^2)/u. \tag{6.62}$$

Clearly we shall have problems if solutions approach the parabola $u = k\dot{u}^2$, or if $u \to 0$.

The first task is therefore to establish whether there is a range of values of ν for which $\ddot{u} < 0$ for u just greater than $k\dot{u}^2$, since this will allow us to demonstrate the existence of a region $0 < \dot{u} < \sqrt{(u/k)}$ which is an attracting set for trajectories of (6.62). Then clearly since $\dot{u}$ is bounded by its initial value or 1, whichever is greater, we find $\dot{u} \to 1$ as $u \to \infty$, as in the case $\nu = 0$. As before we have $\ddot{u} > 0$ for $\dot{u} \to 0+$ (certainly for $u \geq 1$ which is all that concerns us). We therefore want exactly one change of sign of $\ddot{u}$ between $\dot{u} = 0$ and $\dot{u} = \sqrt{(u/k)}$. It can be seen that for $-1/6 < \nu \leq 0$ one branch of the curve $\ddot{u} = 0$ starts at the origin and has $\dot{u}$ increasing monotonically to 1 as $u \to \infty$. Also near the origin the $\ddot{u} = 0$

curve is $\dot{u} = -2\nu u/k + O(u^2)$, which lies in the required region.

Hence it suffices to show that the curve $\ddot{u} = 0$ does not cross $u-k\dot{u}^2 = 0$. The simultaneous solution of these equations requires

$$f(\dot{u}) \equiv \dot{u}(1+2\dot{u})(1-\dot{u}) + (1/2\nu+\dot{u}+2\dot{u}^2) = 0 \tag{6.63}$$

Now $f(0) = 1/2\nu < 0$ and $f(\dot{u}) \sim -2\dot{u}^3 < 0$ for $\dot{u} \to \infty$ so we may have no roots or two roots in $0 \leq \dot{u} < \infty$, and find out which by looking at the sign of $f(\dot{u})$ at its maximum. This is at

$$f(\dot{u}) = -6\dot{u}^2 + 6\dot{u} + 2$$

and the positive root of this is $\dot{u}_m = \frac{1}{2} + \surd(7/12)$. Then $f(\dot{u}_m) \geq 0$ if $\nu > -3/(9+7\surd(7/3)) = \nu_{crit}$ which is approximately -0.152 and restricts ν more than the more obvious constraint $\nu > -1/6$.

Therefore for $\nu_{crit} < \nu < 0$ we have behaviour qualitatively similar to that for $\nu = 0$, provided $\dot{u}(0) < \surd(1/k)$ i.e. $u'(0) < \surd(-1/6\varepsilon\nu\alpha)$. We should note that there are solutions for $u > \surd(u/k)$ which also exist and have both u and $\dot{u}$ increasing and $\surd(u/k) < \dot{u} < ku$. It is not clear whether these 'extra' solutions are physically reasonable or not, and if they are what influence they have on the general qualitative picture of the behaviour of this model. If we recall that ν was introduced as a (small) coefficient to improve predictions of the model, and note that the dominant terms leading to the 'extra' solutions involve ν, we may be disposed to doubt the relevance of these solutions.

Finally, as a relatively simple example of the effect of a larger magnitude of ν, we consider $\nu = -1/6$. In this case $\ddot{u} = 0$ on $\dot{u} = 1$ and on $u = k\dot{u}(3+2\dot{u})/(1+2\dot{u})$ and there is a region on $u - k\dot{u}^2 = 0$ between $(k,1)$ and $(9k/4, 3/2)$ where $\ddot{u} > 0$ for $\dot{u}$ just less than $\surd(u/k)$. Since $\dot{u} = 1$ is a trajectory in this case, solutions with $\dot{u}(0) \leq 1$ and $0 < \dot{u}(0) < \surd(1/k)$ will exist for all x and will have $\dot{u}(x) \to 1$ as $x \to \infty$. Also if $u(0) \geq 9k/4$ (i.e. $k \leq 4/9$) and $0 < \dot{u}(0) < \surd(1/k)$ solutions are confined to the region bounded by the inequalities $u(x) > 9k/4$, $0 < \dot{u}(x) < \surd(u(x)/k)$ for all $x > 0$, and hence exist. We can extend the regions of 'allowable' initial conditions slightly by observing that $\ddot{u} < 0$ for $1 < \dot{u} < \surd(u/k)$ and $u > k\dot{u}(3+2\dot{u})/(1+2\dot{u})$. Figure 6.4 shows this phase plane, and the regions to the right of the curve OABCD is the region of 'allowable' initial conditions. In fact there must be a solution curve connecting A and C and this, rather than the curve ABC, is part of the boundary of the maximal 'allowable' regions. The arrowed line segments are sketched segments of trajectories.

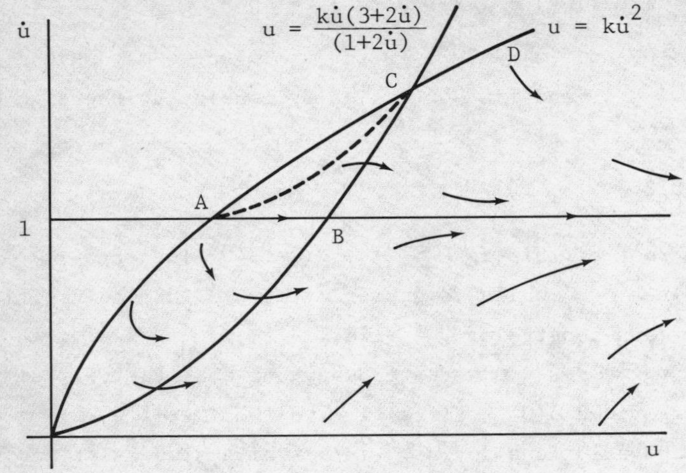

Figure 6.4 *Phase plane for the modified Maxwell model (6.62) for* $\nu = -1/6$.

For values of ν larger in magnitude than $-1/6$ the region of 'allowable' initial conditions is smaller, and all solutions which do exist as $x \to \infty$ will approach $\dot{u} = 1$ from above rather than below as for $\nu < -1/6$. The conclusions here are not that such values of ν are not allowable, but that it is desirable to seek other constitutive equations which do not lead to such restrictions on initial conditions. No physical interpretation of the breakdown of solutions is offered, since it is unlikely that the model is applicable when breakdown occurs. What is, perhaps, remarkable is that the same problem of breakdown of solutions occurs with two models intended to improve the upper convected Maxwell model by allowing shear thinning and a non-zero (negative) second normal stress difference. In both cases we are able to use the parameter describing these improvements only over a limited range of values.

Finally we mention some more recently published work. Fisher and Denn [F10] extend their work (on steady spinning and draw resonance) to nonisothermal flow and present computed results for various values of the Stanton number (a dimensionless heat transfer coefficient). In work on nonisothermal spinning which should be compared with the above Matsui and Bogue [M18] use an inverse of the procedure we have discussed here, taking $u(x)$ from observation and calculating stresses along the spinline (using an integral constitutive equation). These stresses are then compared with measured stresses to test the validity of the constitutive equation. Ronca [R11] uses a recently developed constitutive equation (cf. [P17]), and in passing offers a different formulation of the problem for the upper convected Maxwell model by eliminating kinematic variables and deriving a system of two equations relating quantities similar to our stresses p_{11} and p_{22}. Ronca's constitutive equation is introduced to satisfy a property he calls statistical invariance, and the net result here is that a relation between p_{11} and p_{22} is obtained so that the additional initial condition $T(0)$ is no longer required. The physical meaning of this "loss of memory" is not immediately obvious.

Phan Thien [P22] has used his model (Table 2.5) with a spectrum of relaxation times and has compared predictions with spinning data of Spearot and Metzner [S15] and of Zeichner [Z5]. The comparison is extremely encouraging, and the additional comparison with the upper convected Maxwell model results shows that these are not as successful quantitatively as the work of Denn and Marrucci [D13] might lead one to expect.

6.1.6 The effect of inertia

If we introduce (3.2d) into (3.1), continue to ignore gravity, air drag and surface tension, and make the same geometrical assumptions about the threadline we obtain

$$(p_{11}(x) - p_{22}(X))A(X) = F - \rho Q(U(L) - U(X)), \qquad (6.64)$$

where F is the force applied at the take-up, $X = L$. The appearance of $U(L)$ warns us that we should turn our attention to the boundary-value problem which is the correct model of the physical process, but we shall postpone this (to section 6.2.2). Using $Q = U(X)A(X)$ we obtain

$$p_{11}(X) - p_{22}(X) = CU(X) + \rho U(X)^2 \qquad (6.65a)$$

where

$$C = F/Q - \rho U(L) \qquad (6.65b)$$

and for the initial-value problem C and ρ are given constants, ρ being positive but C possibly negative.

The nonlinearity which this introduces into the mathematical problem has some significant effects, particularly on the global existence of solutions of the initial-value problem. For the Newtonian fluid this is easily seen [M17] since the equation

$$3\eta U'(X) = CU(X) + \rho U(X)^2 \tag{6.66}$$

has the solution

$$U(X) = \frac{CU(0)\exp(CX/3\eta)}{C+\rho U(0)\{1-\exp(CX/3\eta)\}} \tag{6.67}$$

which tends to infinity as $X \to (3\eta/C)\log(1+C/\rho U(0))$. Thus we have a limited interval of existence, determined by the parameters ρ, η, C and the initial value U(0). To have an accelerating flow, as we expect for a positive (tensile) applied force, $C > -\rho U(0)$ so the limiting value of X is real. We shall see below that the correct boundary-value problem always has a solution. The breakdown (or "blow-up") of the solutions of (6.66) at a finite value of X is associated with a situation where the geometrical approximations behind the simple analysis also break down so we should not attempt a detailed physical interpretation (in terms of thread breakage for example). In the sense that $U'(X)/U(X)$ is larger than in the absence of inertia we could say that inertia decreases spinnability [M17], but it is perhaps wisest to leave such discussions until we consider the boundary-value problem (section 6.2.2). The mechanism for the blow-up can be described by saying that when inertia is significant a greater force must be applied at a given cross-section in order to accelerate the fluid. As this greater force is transmitted through the fluid it leads to a correspondingly greater velocity gradient, and this will lead to a yet greater force in a positive feedback loop.

For the Maxwell model two different behaviours are exemplified by the upper convected model and the co-rotational model, and in fact the former is not at all typical. The atypical result, proved by Waters and Petrie [W7], is that for the upper convected Maxwell model, provided that $C > 0$, we have global existence of solutions of the initial-value problem. Furthermore, in contrast to the result in the absence of inertia [D12] there is no limiting value for the strain rate. The more general Maxwell model is discussed elsewhere [P18], and we shall limit ourselves to a few simple observations here. The equations we have to solve are (6.4) and (6.65), and we define dimensionless distance, x, velocity u(x) and relaxation time α as in (6.5b,c), (6.6a). If we postpone defining the characteristic stress (which was FU(0)/Q in (6.5) and (6.6)) and denote this unspecified quantity by S we define parameters

$$K = CU(0)/S, \quad R = \rho U(0)^2/S, \quad G = 3\eta/\lambda S \tag{6.68a,b,c}$$

and the dimensionless variable

$$w(x) = 3ap_{11}(X)/S + G - (1+a)Ku(x) - (2+a)Ru(x)^2 \tag{6.69}$$

which is related to the axial component of the extra stress.

The reason for this choice is that the equations for u(x) and w(x) are

$$\alpha w(x)u'(x) = Ku(x) + Ru(x)^2 \tag{6.70a}$$

and

$$\begin{aligned}\alpha u(x)w'(x) = {} & G - w(x) - (1-a)Ku(x) - (2-a)Ru(x)^2 \\ & + \alpha u'(x)\{(2a-1)(a+1)Ku(x) - 2(1-a)(2+a)Ru(x)^2\}\end{aligned} \tag{6.70b}$$

Then solutions break down if either u(x) or w(x) approach zero for a finite value of x. The parameters α, G and R are positive and $K > -R$ follows from $C > -\rho U(0)$. If $w(0) > 0$ then, since $u(0) = 1$, $u'(0) > 0$ and hence u(x) can only increase while w(x) remains positive. Hence we shall only have existence problems if $w(x) \to 0$, and the proof for $a = 1$, $K > 0$ consists in showing that $w'(x) > 0$ for w(x) close to zero. It may further be shown that

$$w(x) \sim (1-R)\, u(x) \tag{6.71a}$$

and

$$u(x) \sim R^{-1}\exp\{Rx/(1-R)\}. \tag{6.71b}$$

The phase-plane for this system, Figure 6.5, gives an indication of how the proof [W7] proceeds by establishing formally the correctness of the features indicated in the sketch.

For $a < 1$ and $R > 0$ the last term of (6.70b) is non-zero and this makes a qualitative difference to the behaviour of solutions, since for large values of u(x) this will dominate and w'(x) will be negative. Hence a solution will

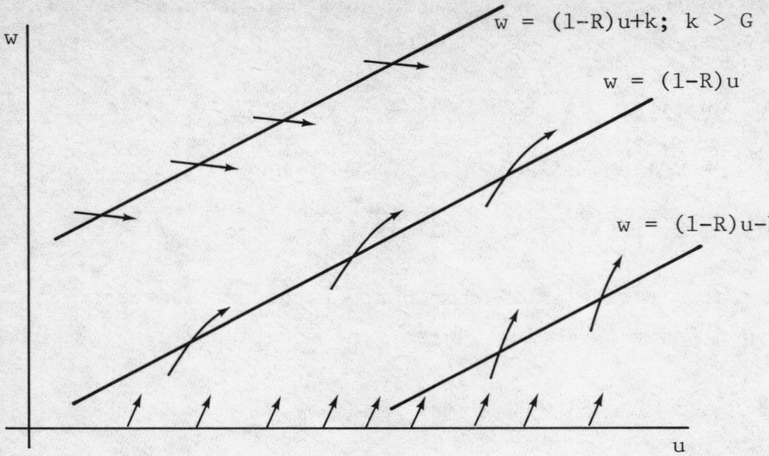

Figure 6.5 *Phase-plane for equations (6.70).*

break down at a finite value of x, with u(x) finite and $u'(x) \to \infty$. This is a different behaviour from that for the Newtonian liquid where $u(x) \to \infty$, but the remarks on interpretation there are still relevant ($u'(x) \to \infty$ implies a violation of the basic assumptions of the geometric approximation).

For $a < \frac{1}{2}$ the sign of $w'(x)$ is negative in most of the quadrant $u > 0$, $w > 0$ of the phase-plane and we get more rapid blow-up of solutions and, as we have seen, this happens even when $R = 0$. The co-rotational model, $a = 0$, has the equations decoupled, so that we may study

$$w(x)u'(x) \equiv (G-Ku(x)-2Ru(x)^2)u'(x) = Ku(x) + Ru(x)^2 \qquad (6.71)$$

alone. It is clear that, for G, K and R such that u(x) increases initially, u(x) will increase until it satisfies

$$2Ru(x)^2 + Ku(x) - G = 0 \qquad (6.72)$$

when $u'(x) \to \infty$, at a finite value of x which could be calculated. This is essentially the same as the behaviour in the absence of inertia.

6.2 BOUNDARY CONDITIONS

The two main problems we discuss here are largely unresolved, but a clear statement of the issues involved is worth some attention. The initial conditions involve a matching of the spinning flow with flow out of an orifice. In addition to this basic unsolved problem we have to consider the influence of the flow history for elasticoviscous liquids. As has been remarked above, the spinning process, and such flows as the ductless siphon and the triple jet, are essentially two-point boundary value problems, where the conditions determining the flow are prescribed in part at the orifice (flow rate, velocity history of deformation) and in part at the take-up or equivalent place (velocity or force applied).

6.2.1 Initial conditions

We have already discussed the way in which specification of initial conditions for a rate equation model and specification of the history of the deformation for an integral equation model are equivalent (section 5.1.2). It is convenient to use the integral formulation to make some very crude estimates of the effect of upstream conditions on the spinning flow. This will be related to the criticism in section 3.3.2 of the idea of separating out the initial stress and its relaxation or decay with time. Following Matsui and Bogue [M18] (see also [P18]) we postulate that the kinematics are given by

$$\underline{C}_t^{-1}(t') = \begin{pmatrix} e^{2\dot{\gamma}(t-t')} & 0 & 0 \\ 0 & e^{-\dot{\gamma}(t-t')} & 0 \\ 0 & 0 & e^{-\dot{\gamma}(t-t')} \end{pmatrix} \tag{6.73a}$$

for $0 < t' < t$ and by

$$\underline{C}_t^{-1}(t') = \begin{pmatrix} (1+\kappa^2 t'^2)e^{2\dot{\gamma}t} & -\kappa t' e^{\dot{\gamma}t/2}(X_2/R) & -\kappa t' e^{\dot{\gamma}t/2}(X_3/R) \\ -\kappa t' e^{\dot{\gamma}t/2}(X_2/R) & e^{-\dot{\gamma}t} & 0 \\ -\kappa t' e^{\dot{\gamma}t/2}(X_3/R) & 0 & e^{-\dot{\gamma}t} \end{pmatrix} \tag{6.73b}$$

for $t' < 0 < t$ (where $R^2 = X_2^2 + X_3^2$). This corresponds to uniaxial elongation with constant strain rate $\dot{\gamma}$ for $t' > 0$ and to axisymmetric simple shear with constant shear rate κ for $t' < 0$. Neither the shear flow nor the transition are consistent with the equations of motion and $\dot{\gamma}$ is not normally constant in spinning, but we ignore these important facts in the hope that, while we will not obtain a solution to the dynamical problem, we may at least gain some insight into the transition for a simple constitutive equation. We use the rubberlike liquid (2.26) with one relaxation time and obtain the extra-stresses

$$p_{11}(t) = \frac{2G\lambda\dot{\gamma}}{1-2\lambda\dot{\gamma}} + \left(2\lambda^2\kappa^2+1 - \frac{2\lambda\dot{\gamma}}{1-2\lambda\dot{\gamma}}\right) Ge^{-(1-2\lambda\dot{\gamma})t/\lambda} \tag{6.74a}$$

$$p_{22}(t) = p_{33}(t) = \frac{-G\lambda\dot{\gamma}}{1+\lambda\dot{\gamma}} + \left(1 + \frac{\lambda\dot{\gamma}}{1+\lambda\dot{\gamma}}\right) Ge^{-(1+\lambda\dot{\gamma})t/\lambda} \tag{7.74b}$$

and

$$(R/X_2)p_{12}(t) = (R/X_3)p_{13}(t) = G\lambda\kappa e^{-(1-\frac{1}{2}\lambda\dot{\gamma})t/\lambda} . \tag{6.74c}$$

Now our basic assumption in analysing the elongational flow is that $p_{12} = p_{13} = 0$, so we see that if we assume this we need to be far enough downstream of the orifice for (6.74c) to have decayed to zero. In that case (6.74b) will also have decayed to zero, and this may be the best clue we can get to the correct choice of initial stress $p_{11}(0)$ (or dimensionless T(0)) at the origin of our X coordinate in the analysis in section 6.1. In the absence of inertia $p_{11}(x) - p_{22}(X) = F\,U(X)/Q$ and so if we take $p_{22}(0) = 0$ then we obtain $p_{11}(0) = FU(0)/Q$ or, (6.5a), $T(0) = 1$.

This approach, of course, does not begin to tackle the real problem of extrudate swell, and the solution of the problem of what the flow is at and near the orifice from which the jet emerges. There is direct experimental evidence on two relevant aspects of this, that changes in the flow upstream of the orifice can have a significant influence on the spinning flow [O7], and that the applied force at the take-up may influence appreciably the amount of extrudate swell [W14]. These results emphasise the importance of the deformation history of the liquid and the fact that we have a two-point boundary-value problem. Furthermore, for the purely elongational flow the initial condition (after the extrudate swell region) is influenced by the downstream boundary condition.

Denn and Fisher and, independently, Caswell and collaborators report that they are each currently working on the computational solution of this problem. The approach involves the use of finite element techniques for the flow within a capillary, through the transitional (extrudate swell) region and into the spinning region. The downstream boundary condition is apparently different in the two cases. Neither of these pieces of work has been published yet.

We shall not discuss extrudate swell here, but direct the reader to some recent papers which lead into the considerable literature on the subject. Detailed experimental studies of the kinematics are providing increasingly useful data [C21,G18] to compare with numerical solutions of the problem, which are as yet confined to Newtonian liquids [N4] although recently results for the second-order fluid are said to have been obtained by Tanner. Experimental data on die swell also continues to be published, together with correlations based on simple theoretical arguments [L14,V11], but these should be read in conjunction with more careful theoretical discussion [D2,J10]. In particular it must be emphasised that the transverse velocity profile in fully developed flow in a tube begins to change before the tube exit, so that momentum balances based on the assumption of fully developed flow at the tube exit (or orifice from which the jet emerges) are in error [D2]. There is additional evidence from flow birefringence studies on this point [V6].

6.2.2 The boundary-value problem

In practice, the spinning process is run with a given flow rate of material from a given orifice, which determines the initial velocity U(0) and the parameter Q, and a given take-up velocity. This emphasises the structure of the flow as a two-point boundary-value problem which it is whether we prescribe the take-up force or the take-up velocity. If we can neglect inertia then

$$FU(X)/Q = p_{11}(X) - p_{22}(X) \tag{6.1}$$

is in effect a first integral of the equation of motion (local conservation of axial momentum) and if F is given explicitly we are left with the initial-value problem discussed above. When the take-up velocity is given we have to find what value of F gives the solution satisfying

$$U(L) = D_R U(0) \tag{6.75a}$$

or, in dimensionless terms

$$u(1) = D_R, \tag{6.75b}$$

where D_R is called the draw ratio.

In the Newtonian case, for example, we have, from (2.98)

$$U(L) = U(0) \exp(FL/3\eta Q)$$

and hence

$$F = (3\eta Q/L) \log D_R. \tag{2.100}$$

Thus there is a one-to-one relation between F, in the interval $(0,\infty)$, and D_R, in the interval $(1,\infty)$. In the case of the upper convected Maxwell model there is a more restricted range of solutions. The initial-value problem has a solution for any positive value of F, but the corresponding range of values of D_R is from 1 to $1 + \alpha^{-1}$ so the boundary-value problem does not have a solution for larger values of D_R.

When we include the effects of inertia things get a little more interesting and in the Newtonian case the result is perhaps a little unexpected. The limited interval of existence of solutions for the initial-value problem (6.66) no longer causes trouble and we have solutions for any positive applied force or any draw ratio greater than one. To prove this [P18] we note that C in (6.66) is determined either by F or by U(L) through (6.65b) and the solution (6.67). Putting these together gives, writing $k = C/\rho U(0)$ and $\xi = \rho QU(0)L/3\eta$,

$$D_R = ke^{k\xi}/\{k + 1 - e^{k\xi}\} \tag{6.76a}$$

and

$$F/\rho QU(0) = k + D_R \tag{6.76b}$$

This determines D_R for any k except zero. If $k = 0$ we have to solve (6.66) with $C = 0$ and the equivalent of (6.76a) is

$$D_R = 1/(1 - \xi) \tag{6.76c}$$

For $k = -1$ we find $D_R = 1$, which is obvious since if $k = -1$ the right-hand side of (6.66) is zero for $U(X) = U(0)$ and so this is the solution of the differential equation, and we have no stretching, and not surprisingly (6.76b) gives $F = 0$. For $k > -1$, $D_R > 1$, provided that k is less than the value which makes the denominator of (6.76a) zero. When k reaches this value, i.e.

$$k + 1 = e^{k\xi} \tag{6.77}$$

D_R becomes infinite, and so does F. Thus we may argue that the limited interval of existence of solutions for the initial-value problems corresponds to assigning values of k which are not physically realistic. Any draw ratio may be obtained by applying a suitable force, and any force may be applied and will determine a corresponding draw ratio. The value of k (or C) determined by this gives an interval of existence greater than L for solutions of the initial value problem.

The discussion of the boundary-value problem for Maxwell models when inertia is important is even more complicated [P18] and no general conclusions have been obtained. It appears that the co-rotational model has solutions only for a limited range of draw-ratios, and that there are restrictions on the force which may be applied. On the other hand the upper convected Maxwell model allows any force to be applied and inertia extends the range of attainable draw ratios.

6.3 MECHANICS OF NEWTONIAN JETS

In the previous sections we have devoted a good deal of effort to the study of the consequences of the choice of constitutive equation on the mathematical model of spinning, and this is one of the main types of investigation appropriate to non-Newtonian fluid mechanics. However the basic model is somewhat simplified and its validity rests on a number of approximations and idealizations which merit further examination. There is no mathematical work concerning this for elasticoviscous fluids (except insofar as some results are independent of the choice of constitutive equation) and we shall discuss now the limited amount of relevant work on viscous liquids. The two topics we shall study are the effect of gravity, in the course of which we make allowance for surface tension and inertia, and the underlying geometrical assumption that the cross-section of the jet changes slowly, with the kinematical consequence that the velocity is sensibly constant across any cross-section. By confining attention to Newtonian liquids some of the techniques of modern (Newtonian) fluid dynamics can be brought to bear, but it must be said immediately that a number of problems remain unresolved and others merit reexamination.

6.3.1 Viscous jets under gravity

We consider an axisymmetric jet of a Newtonian liquid under the influence of gravity, inertia, surface tension and an externally applied take-up force.

Then (3.1), with the omission of air drag, is

$$T_X A(X) = F + F_{gravity} - F_{surface} - F_{inertia}$$

where F is the take-up force, applied at X = L, and T_X is the stress at the cross-section. From the discussion following that equation we obtain, using (3.2a), (3.2c), (3.2d) and (3.6) (as well as the Newtonian constitutive equation),

$$3\eta U'(X)A(X) = F + \int_X^L \rho g A(x)dx - \pi\sigma(R(X)-R(L)) - \rho Q(U(L)-U(X)). \quad (6.78)$$

Differentiating this and using $A(X) = Q/U(X)$ and $R(X) = \sqrt{(A(X)/\pi)}$ gives the second order equation

$$3\eta(U''(X) - U'(X)^2/U(X))$$
$$= -\rho g + \tfrac{1}{2}\sigma\sqrt{(\pi/Q)}U'(X)/\sqrt{U(X)} + \rho U(X)U'(X). \quad (6.79)$$

This can be solved for various special cases, as has been done by Trouton [T24] (viscous terms and gravity), Marshall and Pigford [M15] (gravity and viscous terms, and gravity and inertia), Matovich and Pearson (viscous terms with gravity, with inertia, and with surface tension) [M17].

Clarke [C19] has considered the effect of viscosity and surface tension on a jet dominated by inertia and gravity, having shown that for a long enough jet the inertia term will eventually grow to be the major one balancing gravity. In the limit the liquid in the jet behaves essentially as though it were falling freely at a velocity

$$U(X) = \sqrt{\{2g(X - \xi)\}} \quad (6.80)$$

where ξ is a constant. As the jet accelerates, it gets thinner and surface tension is bound to become important eventually. When it does it leads to the pressure inside the jet being $\sigma/R(X)$, which increases without bound (until the jet breaks). In the absence of surface tension Clarke also shows that the pressure tends to zero (the ambient pressure outside the jet is taken to be zero) from below. The solution obtained by Matovich for the case where surface tension and viscous forces alone are considered shows that the effect of surface tension becomes negligible at large X, when U(X) is large, but of course this is the situation where inertia becomes important so that this deduction cannot be supported by a rigorous analysis (as Matovich clearly indicates) and Clarke's conclusion must be the correct one.

Neglecting the effect of surface tension (as we shall do from now on, having noted that this cannot safely be done unless we check afterwards on the magnitude of the neglected forces) Kaye and Vale [K13] were able to predict the shape of a jet of a silicone fluid of kinematic viscosity (η/ρ) of 300 stokes (.03 $m^2 s^{-1}$) with good agreement if the radius and its derivative at one point were used to start the numerical solution. This simple theory was less successful with a liquid of kinematic viscosity 10 stokes, and a theory involving the variation of velocity across the cross-section of the jet was developed (see 6.3.2 below).

In the case where viscosity and gravity are the dominant effects, the equation is

$$U''(X) - U'(X)^2/U(X) \quad = \quad -\rho g/3\eta \tag{6.81}$$

and this has the solutions

$$U(X) \quad = \quad (\rho g/6\eta A^2)\ \sinh^2(A(X+B)), \tag{6.82a}$$

$$U(X) \quad = \quad (\rho g/6\eta)(X+B)^2 \tag{6.82b}$$

and

$$U(X) \quad = \quad (\rho g/6\eta A^2)\ \sin^2(A(X+B)), \tag{6.82c}$$

where A and B are arbitrary constants to be determined by the boundary conditions. Trouton quotes an expression for the radius equivalent to (6.82a) with B = 0 as the general solution, and also (in the limiting case $A \rightarrow 0$) (6.82b) which gives a hyperbolic shape for the jet,

$$XR(X) \quad = \quad \surd(6\eta Q/\pi\rho g). \tag{6.83}$$

Matovich and Pearson obtain the general form (6.82a), which they attribute to Trouton (arguably the arbitrary choice of origin for X allows B to be taken as zero) and Marshall and Pigford, because of a sign error in their equation, obtain a function which is not a solution to the given problem, although its behaviour is plausible.

We should note that the initial conditions taken by Marshall and Pigford are $U(X) = U_o$ and $U'(0) = 0$, from the physical hypothesis that the jet emerges from the orifice with parallel sides. These boundary conditions lead to solutions of the form (6.82c) for the correct equation, (6.81), and

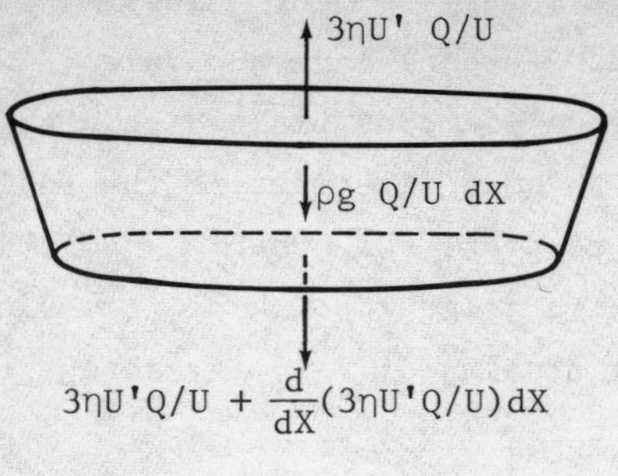

Figure 6.6 *Balance of gravity and viscous forces in a jet.*

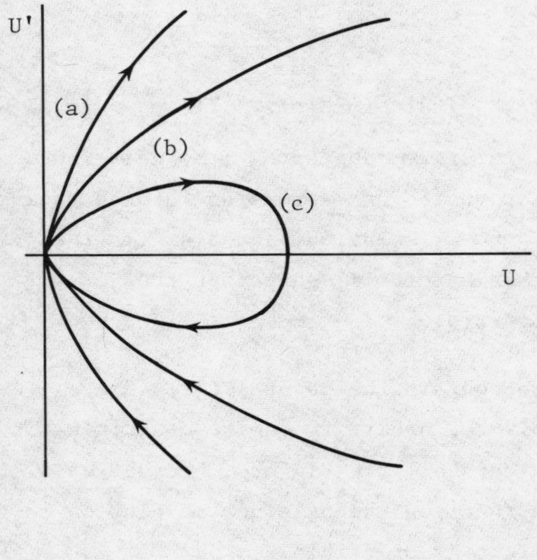

Figure 6.7 *Phase-plane for equation (6.81).*

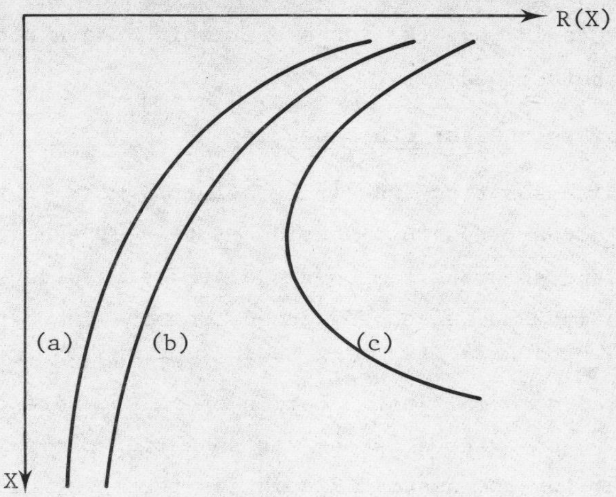

Figure 6.8 *Jet shapes determined by gravity and viscosity.*

to a jet which is not stretched. We may interpret this physically as a situation where the weight of material below the orifice is not enough to stretch the jet. Note that the simple force balance for this situation (Figure 6.6) shows clearly that the viscous force $3\eta U'(X)Q/U(X)$ decreases as we go down the thread, because there is less weight to support further down the thread. The solutions (6.82) can be represented as graphs of $U(X)$, but it may be more helpful to use the phase-plane representation (Figure 6.7). We have the explicit expression

$$U'(X)^2 = 4U(X)(\rho g/6\eta + CU(X)) \qquad (6.84)$$

for the trajectories, with $C = A^2, 0, -A^2$ giving (6.82a), (6.82b) and (6.82c) respectively. The corresponding solutions are more easily visualised if we plot $R(X) = \sqrt{(Q/\pi U(X))}$ rather than $U(X)$, and typical jet shapes are sketched in Figure 6.8.

A calculation which ought to be done is the correction of that done by Marshall and Pigford, which showed for the wrong equations how well the neglect of inertia initially and of viscosity finally approximated the jet shape obtained by numerical solution of the full equation with all three terms in it, i.e. (6.79) without surface tension. It is expected that their result would be reproduced, but the range over which the approximations are useful has not been determined.

6.3.2 Variations across the jet cross-section

The basic kinematical assumption underlying all this work is that the axial velocity is uniform across the cross-section of the jet and that other velocity components and shearing components of the stress tensor can be neglected. This can in some sense be deduced from the geometrical assumption (or observation) that the rate of change of the jet radius is small, but it is not easy to provide a formal justification of this. Note that the free surface will only have an external shear stress acting on it if air drag is significant, and that for the assumed kinematics the free surface does have an internal shear stress acting on it. The shear-free surfaces in the liquid are parallel to the Cartesian coordinate axes, and the assumption of the idealised kinematics is thus clearly an approximation. It does not seem likely that air drag could provide a shear stress which varied in the correct way. We shall discuss briefly here four investigations into aspects of this problem, and refer readers to the original papers for details.

Matovich and Pearson [M17] consider the full equation of motion and postulate a series expansion for each variable whose leading (zero-order) terms embody the basic assumptions we have listed above. The succeeding terms in the series are, "in some sense related to the smallness of" $R'(X)$, smaller than the preceding terms. There is no explicit small parameter, but it is possible (at least for a Newtonian liquid) to show that the first-order terms are of order $R'(X)$ compared with the zero-order terms. The difficulties that this approach meets are set out clearly by Matovich and Pearson and in particular they note that a scheme for actually calculating the first-order terms was not obtained. They suggest that the use of streamline coordinates would assist in this, and Clarke in fact did this in a paper published a few months earlier [C19].

The reformulation makes it easier to deal with the boundary conditions at the free surface, at the expense of some additional complexities in the equations. The correct conditions are that the velocity vector of the fluid lies in the surface and that the stress in the fluid at the free boundary is equal to that in the surrounding air (an isotropic pressure, taken to be zero). Of course this is modified if surface tension is taken into account, as in the work of Clarke (and of others, [M17,K13]). The problem of finding a small parameter is no easier and Clarke uses the same sort of technique as Matovich, introducing an artificial dimensionless parameter. He is also forced to obtain a zero-order solution by making "the simplifying assumptions that the jet will be very thin and that variations across the jet will be small", at least far downstream. The more rigorous treatment here is made possible by looking explicitly for an asymptotic solution valid far downstream, which we have mentioned above (section 6.3.1). The solution does also contain terms which provide "an estimate of the rate of decay to a uniform velocity profile", although this is not explicitly worked out. The difficulty lies in finding an appropriate upstream boundary condition for this asymptotic solution. In the case of two-dimensional flow (a sheet rather than a jet) Clarke [C18] is able to use complex variable techniques to obtain an "inner solution" to match with the far downstream "outer solution". Even this "inner solution" is valid only in a region away from the orifice, but shows how the full Navier-Stokes equations may be treated to obtain valid approximate solutions.

Kaye and Vale [K13] attempt to solve the same problem using the stream function ψ as a dependent variable, and developing series solutions in the form

$$\psi(X,r) = \sum_{k=1}^{\infty} r^{2k} q_k(X), \tag{6.85a}$$

$$q_k(X) = \sum_{j=1}^{\infty} \frac{\alpha_{kj}}{X^{a_{kj}}} \tag{6.85b}$$

They take the free boundary to be at $r = R(X)$ and assume

$$R(X) = \sum_{i=1}^{\infty} \frac{K_i}{X^{b_i}} \tag{6.86}$$

The results obtained (ignoring surface tension terms which are in [K13]) are

$$\psi(X,r) = r^2\sqrt{(gX/2)} - (r^4/8)\sqrt{(g/2X^3)} + r^6(5\rho g/512\eta X^2) + \ldots, \quad (6.87a)$$

$$R(X) = K_1/X^{\frac{1}{4}} + \ldots, \quad (6.87b)$$

$$K_1^2 = Q/(\pi\sqrt{(2g)}). \quad (6.87c)$$

This solution is found to be in qualitative agreement with results for their less viscous fluid where the simple theory was inadequate. The surface tension terms which they include in their solution dominate the viscous terms, as is expected for solutions valid for large X, and this is in agreement with Clarke's results. The effect of viscosity on R(X) is not given.

Finally Kase [K11] has attempted to study the more practical problem where there is a temperature gradient both across the jet and along it (as would be the case in melt spinning) and a consequent variation in viscosity. He assumes a velocity profile of the form

$$U = U_o e^{\alpha X}(1 + a_2 r^2 + a_4 r^4 + \ldots) \quad (6.88a)$$

where the parameters are functions of X, and discusses numerical examples based on two possible sets of spinning conditions. Even in the unfavourable case of a very short thread the nonuniformity of velocity is very small;

$$a_2 R^2 + a_4 R^4 = -1.43 \times 10^{-3}. \quad (6.88b)$$

None of this leads to a wholly satisfactory mathematical conclusion. The recourse to the "far downstream" asymptote is not a great deal of help in dealing with the boundary-value problem of spinning, and in fact any jet falling under gravity must be influenced by a downstream boundary condition in the same way. However the simple model of the earlier sections has not been destroyed by any careful theoretical attack and there is no reason at all to suppose that it is not an adequate and correct approximation. What would be helpful would be an indication of when the approximation breaks down and how to deal with such situations. The work described above offers some help in this. It is interesting to see that the apparently much more complicated film-blowing process can be dealt with more easily [P4], just because there is a natural small parameter (film thickness/radius of curvature). The approach in that work also uses curvilinear coordinates which are essentially the same as the streamline and orthogonal coordinates of Clarke.

6.3.3 A Cosserat-type director theory

Green [G22], following Green and Laws [G21] has approached the jet problem in a totally different way. The earlier work is concerned with ideal (inviscid) fluids and in the later work viscosity is considered. The starting point is the development of an explicitly one-dimensional theory for the jet, with coordinate measured along the jet (which is taken to lie along a curve in three-dimensional space). The jet is given some structure by defining two vectors, called directors, at each point on the curve. Equations of motion for a point on the curve and for the directors at this point must be set up, and constitutive equations are also required. The result is a somewhat formal abstract theory, and in order to use this we need to connect it with our knowledge of mechanics and material response in three dimensions. Green makes this connection and in particular recovers our equations (6.79) for a straight circular jet if he sets the ratio of axial to radial length scales equal to zero. There is then the suggestion that when this ratio is non-zero the theory takes account of variations across the jet, and indeed Bogy [B23] claims that the theory includes terms associated with transverse (radial) inertia.

It is not clear at this stage in its development whether this theory offers an adequate return for the effort involved in its construction and connection with conventional theories. In particular it appears that it may not help in investigating the behaviour of elasticoviscous fluids in elongational flows and in comparing this with behaviour in shearing flows. The theory, if it succeeds at all, will be appropriate to jets only - a modified theory (and different material parameters) will be required even for flow of a sheet, although this is also an elongational flow. There is no doubt on the one hand that the theory offers some interesting mathematical problems and on the other hand that its practical use will meet considerable difficulties. Naghdi [N1] has used these ideas in a recent study of the temporal stability of a jet and compares results with the classical work of Rayleigh and Weber. This is complementary to Bogy's spatial stability analysis [B23].

6.4 STABILITY

We have already (sections 1.2.3) discussed this topic briefly and cited several recent review papers on stability of shearing and elongational flows of polymeric liquids. Here we shall present the equations for unsteady flow of threads or jets of liquid and discuss two aspects in a little detail, namely linearised stability theory and some features of the full nonlinear partial differential equations.

We continue to assume that the velocity is uniform across the cross-section of the thread, and neglect gravity, inertia and surface tension. This means that we are looking at viscous flow and hope to see whether necking or possibly periodic fluctuations in diameter will occur. The equations governing the flow are obtained either by straightforward mass and momentum balances on the thread as a whole, or by integrating the Navier-Stokes equations (or equivalent for an elasticoviscous liquid) across the cross-section. The result is the continuity equation

$$(UA)_x = -A_t \tag{6.89}$$

and the momentum equation

$$(TA)_x = 0 \tag{6.90a}$$

where the subscripts here denote partial derivatives, and for simplicity the dependence of velocity U, cross-sectional area A and stress T on the axial coordinate X and on time t is not explicitly indicated. The constitutive equation completes the model, and the time-dependence of the dependent variables will have an explicit effect here for elasticoviscous models. Suitable initial and boundary conditions are the functions $U(0,t)$, $A(0,t)$ for $t > 0$, $U(X,0)$ and $A(X,0)$ for $0 < X < L$, and one condition at the take-up - either the velocity $U(L,t)$ or the force $F(t) = T(L,t)\,A(L,t)$ for $t > 0$.

6.4.1 Newtonian thread: nonlinear stability

We have $T = 3\eta U_x$, and hence (6.90a) becomes

$$3\eta(U_xA)_x = 0 \tag{6.90b}$$

which may be integrated to give

$$3\eta U_xA = F(t). \tag{6.91}$$

Here we see that the applied force F may be a function of time, and this will be prescribed explicitly or else determined implicitly by prescribing the value of the take-up velocity $U(L,t)$. We have then, if $F(t)$ is prescribed, a pair of first-order quasilinear equations [J2] which turn out to be hyperbolic but have a somewhat unusual structure. We write the system

$$\begin{pmatrix} 0 & 0 \\ 0 & 1 \end{pmatrix} \begin{pmatrix} U_t \\ A_t \end{pmatrix} + \begin{pmatrix} A & 0 \\ A & U \end{pmatrix} \begin{pmatrix} U_x \\ A_x \end{pmatrix} = \begin{pmatrix} F/3\eta \\ 0 \end{pmatrix} \tag{6.92}$$

and see that this is not in divergence form and neither is it hyperbolic in the t-direction. However it is strictly hyperbolic [J2, p.45] in the X-direction, and can be written

$$\begin{pmatrix} U_x \\ A_x \end{pmatrix} + \begin{pmatrix} 0 & 0 \\ 0 & U^{-1} \end{pmatrix} \begin{pmatrix} U_t \\ A_t \end{pmatrix} = \begin{pmatrix} F/3\eta A \\ -F/3\eta U \end{pmatrix} \tag{6.93}$$

This has characteristics $dt/dX = 0$ and $dt/dX = U^{-1}$ and along the first set of these $dU/dX = F/3\eta A$, while along the second set $dA/dX = -F/3\eta U$. If we have solutions with U increasing along the thread (as X increases) at all times, then the characteristics may be sketched as in Figure 6.9 and the second set ($dt/dX = U^{-1}$) diverge from each other, so that solutions will be well-behaved. It is not clear how to deal with the boundary-value problem where F(t) is determined to give the specified U(L,t), but it does not, at first sight appear to be properly posed since it involves giving data on two "time-like" arcs [J2, p.67], X = 0 and X = L.

If we write the system as a third-order one, using (6.90b) instead of

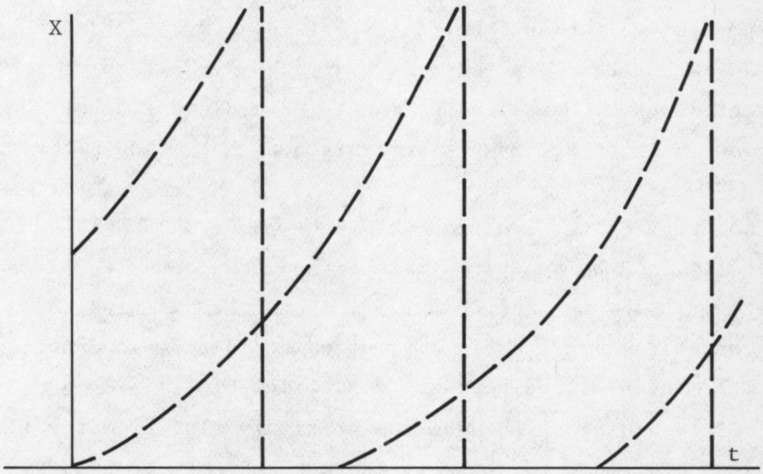

Figure 6.9 Characteristics for the system (6.93).

(6.91), we have corresponding to (6.93) the system

$$\begin{pmatrix} U_x \\ T_x \\ A_x \end{pmatrix} + \begin{pmatrix} 0 & 0 & 0 \\ 0 & 0 & -T/AU \\ 0 & 0 & 1/U \end{pmatrix} \begin{pmatrix} U_t \\ T_t \\ A_t \end{pmatrix} = (3\eta)^{-1} \begin{pmatrix} T \\ T^2/U \\ -AT/U \end{pmatrix} \tag{6.94}$$

which is hyperbolic, with characteristics $dt/dX = 0$ and $dt/dX = U^{-1}$. Along the first of these we find (not surprisingly) $dU/dX = T/3\eta$ and $dT/dX = T^2/3\eta U$, while along the second $dA/dX = -AT/3\eta U$. This formulation of the problem has not yet helped in the analysis of (6.89) and (6.90), and there are clearly some interesting mathematical problems here.

Hyun [H20] has attempted a similar formulation, in terms of nonlinear wave propagation, in which he discusses throughput waves (the quantity propagated is the throughput or volume flow rate, UA). Unfortunately, as Denn [D15] points out there is an error in logic in Hyun's development which leads to a contradiction and invalidates his analysis completely. Hyun obtains a critical draw ratio of 19.744 for a Newtonian liquid, where the accepted value of 20.218 has been obtained by a number of methods and is undoubtedly correct (to at least three significant figures) for stability to small perturbations. This fact alone should persuade even the non-mathematician that all is not well, and Denn demonstrates convincingly that the error lies in the manipulation of partial derivatives associated with changing independent variables from (X,t) to (X,A). Hyun "deduces" that U is linear in X (rather than exponentially varying (2.98)), which again is clearly in error. Hence, whatever the merits of a discussion of nonlinear wave propagation in spinning, Hyun's analysis does not lead to any trustworthy conclusions.

Numerical solution of the equations has been undertaken by Ishihara and Kase [I6] directly, and by an approximate method developed from linearised stability analysis by Fisher and Denn [F8]. Both methods give the same result, that there is a limit cycle corresponding to sustained periodic oscillations of U and A if the boundary conditions $U(0,t) = U_o$ and $U(L,t) = U_oD_R$ are prescribed and if D_R exceeds the critical value from the linearised analysis. This corresponds to the well-known phenomenon of draw resonance which Donnelly and Weinberger [D24] have measured carefully for a silicone oil. Chang and Denn [C9] have similar measurements, and imply that the silicone oil used by Donnelly and Weinberger is not Newtonian, while their corn syrup is.

6.4.2 Linearised stability analysis

In view of the difficulty of solving the nonlinear equations (6.89) and (6.90) or (6.91) in general, it is profitable to consider solutions close to the steady solutions we have obtained earlier. This allows us to decide whether such solutions are stable to small distrubances, although it does not determine the ultimate fate of a solution which is unstable, nor does it provide definite answers to questions about stability to finite disturbances.

The steady solution for the Newtonian liquid is

$$U(x,t) = U_s(x) \equiv U_o \exp(F_s X/3\eta Q) \tag{6.95a}$$

$$A(x,t) = A_s(x) \equiv A_o \exp(-F_s X/3\eta Q) \tag{6.95b}$$

$$U_o A_o = Q \tag{6.95c}$$

and we seek a solution of the form

$$U(x,t) = U_s(x)(1+\tilde{u}(x,t)) \tag{6.96a}$$

$$A(x,t) = A_s(x)(1+\tilde{a}(x,t)) \tag{6.96b}$$

where $\tilde{u}$ and $\tilde{a}$ represent small perturbations to the steady flow. We substitute (6.96) into (6.89) and (6.90), and simplify the resulting expressions by linearising in $\tilde{u}$ and $\tilde{a}$ and also by using the fact that $U_s(X)$ and $A_s(X)$ are solutions. The final linearised stability equations are

$$U_s(X)(\tilde{u}_x+\tilde{a}_x) = -\tilde{a}_t \tag{6.97a}$$

$$U_s(X)\tilde{u}_x + U_s'(X)(\tilde{u}+\tilde{a}) = F_s\tilde{f}/3\eta A_s \tag{6.97b}$$

where the axial force has been written

$$F(t) = F_s(1+\tilde{f}(t)) \tag{6.96c}$$

where F_s is the constant force giving the steady solution (6.95). Boundary conditions for this are

$$\tilde{u}(0,t) = \tilde{a}(0,t) = 0 \tag{6.98a}$$

and either

$$\tilde{u}(L,t) = 0 \tag{6.98b}$$

if the take-up velocity is specified or

$$\tilde{f}(t) = 0 \tag{6.98c}$$

if the take-up force is specified. In posing the problem with these boundary conditions we are asking whether the steady flow is stable to an infinitesimal disturbance anywhere on the thread. It is of equal importance to ask if the steady flow is stable to disturbances to the boundary conditions, but in the linear theory the answer is the same. We have a homogeneous problem with homogeneous boundary conditions for the perturbations $\tilde{u}$ and $\tilde{a}$ and hence seek nontrivial solutions or eigen-solutions. We do this by a spectral decomposition of the perturbations, writing

$$\tilde{u}(X,t) = \sum_k \beta_k(X) \exp(\lambda_k t) \tag{6.99a}$$

$$\tilde{a}(X,t) = \sum_k \alpha_k(X) \exp(\lambda_k t) \tag{6.99b}$$

and, since the problem is linear, studying the behaviour of a typical mode. We seek a mode for which λ_k has a positive real part, since this corresponds to a perturbation which will grow with time and hence indicates that we have an unstable steady flow. If we take boundary condition (6.98b) and write

$$\tilde{f}(t) = \phi \exp(\lambda t) \tag{6.99c}$$

(with ϕ constant) corresponding to the mode λ with eigenfunctions $\alpha(X)$ and $\beta(X)$, then ϕ turns out to be an arbitrary magnitude of the disturbance and the solution is

$$\alpha(X)/\phi = -F(U(X)/U_o,\lambda) \tag{6.100a}$$

$$\beta(X)/\phi = 1 - \exp\{\lambda(U(X)/U_o-1)\} + \{1-\lambda U_o/U(X)\}F(U(X)/U_o,\lambda) \tag{6.100b}$$

where

$$F(u,\lambda) = \int_1^u x^{-1} \exp\{\lambda(x-1)\}dx \tag{6.100c}$$

and the eigenvalue λ satisfies

$$0 = 1 - \exp\{\lambda D_R-1\} + \{1-\lambda D_R^{-1}\}F(D_R,\lambda). \tag{6.100d}$$

Numerical solution of (6.100d) shows that all roots λ have negative real parts if $D_R < 20.218$, and that at least one root has a positive real part for $D_R > 20.218$. The critical eigenvalue is complex and hence predicts an oscillatory breakdown of the steady flow. This has been well substantiated by experiments, as has been stated above [D24,I6,C9], although clearly the large fluctuations observed are not described by the linearised theory. Extensions of this work to elasticoviscous liquids are described elsewhere [F10] and reviews [P14,P5,W18,I2,I3] have been cited. Denn [D11] and Joseph [J11] have included discussion of elasticoviscous liquids in their books on stability although only Denn deals with elongational flows.

One final point of considerable importance concerns the boundary conditions. The choices made here in (6.98) are the simplest to discuss but are not the only possible ones. Pearson and Matovich [P3] discuss a variety of possibilities, although a comparison with their work is complicated by their introduction of small perturbations in the boundary condition and study of amplification factors rather than direct study of eigenvalues. Ronca [R11] considers in some detail how the influence of the "neck zone" (which we have called the extrudate swell region) can be accommodated in a stability analysis even though it cannot itself be analysed fully. It is not possible to discover directly what influence this has on solutions of the equations or on conclusions drawn from them since Ronca does not solve the problem for a Newtonian liquid, but just for two elasticoviscous models.

7 Concluding remarks

"now it is known that it is a cosmos and that the intimate laws which govern it have been formulated, at least provisionally."

Jorge Luis Borges

There are two important problems which await satisfactory resolution and which involve elongational flow. We have already discussed the phenomenon of extrudate swell (section 1.2.2) and the associated problem (section 6.2.1) of posing the correct initial conditions for spinning. This is essentially a problem concerning the transition from shearing to elongational flow (and from a confined flow to a free-surface flow). The significance of this problem is far greater than the reader might suppose from the brief mention it has had in these sections and, as we have remarked, much effort is being devoted to making progress both in understanding the mechanics and in performing calculations to estimate the swelling.

The second major problem concerns flows which are intermediate between shearing and elongation in some sense. We shall discuss briefly, as a simple example of this, converging flow. More general ideas of nearly extensional flows are also occupying a number of researchers at present. These should, in time, offer a foundation for more rigorous treatment of converging flow. This type of flow clearly has considerable practical importance as well as fundamental significance. We shall review a number of similar flows where the elongational response of polymeric liquids plays an important part.

7.1 CONVERGING FLOW

We consider two experimental arrangements here. The flow into an orifice, or upstream of an abrupt contraction from a reservoir or large diameter tube to a capillary is termed free convergence, while a gradual, tapered, transition between a large diameter and a small diameter tube gives constrained convergence. In experiments axisymmetric flow through circular tubes is generally simpler, but two-dimensional flow into a slit can be arranged and this is slightly easier to illustrate and discuss.

There is much experimental evidence concerning flow patterns in free convergence, of which two photographs by Giesekus, reproduced by Metzner,

Uebler and Chan Man Fong [M26], show with great clarity typical behaviour of a Newtonian and a polymeric liquid. The former has straight streamlines, and appears to be very like sink flow, sketched in Figure 7.1(a). The polyacrylamide solution has a central region with curved streamlines, similar to those sketched in Figure 7.1(b), with a recirculating eddy on either side. This latter flow is often called wine-glass-stem flow and is observed for many polymer solutions and melts. There are polymeric liquids (e.g. molten high density polyethylene) for which the eddies are absent (or extremely small), and it must also be stated that visual evidence does not give a clear indication of whether the streamlines are better approximated by the straight lines of Figure 7.1(a) or the hyperbolae of Figure 7.1(b).

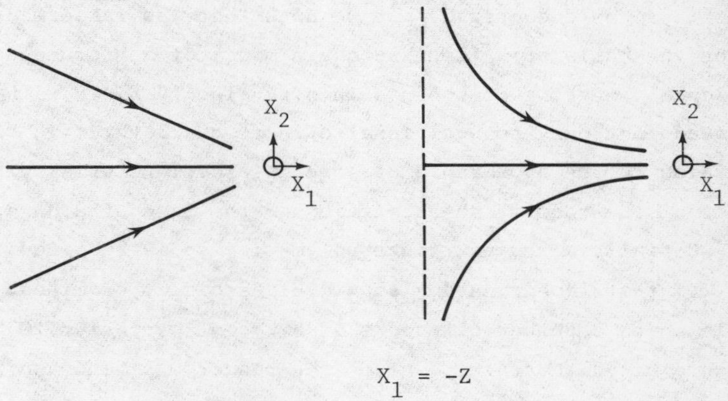

(a) Sink flow *(b) Pure shear*

Figure 7.1 Idealised plane converging flows.

Recent experimental work by Nguyen and Boger [N3] seeks to identify the separate influences of fluid elasticity, shear-thinning behaviour, and inertia. The size of the eddies and the conditions under which steady flow breaks down are studied but, as yet, the dynamics have not received such careful attention. Numerical simulation has been undertaken by a number of researchers [C37,P9] and most recently Crochet and Bezy [C39] have succeeded in predicting the recirculating eddies with a Maxwell model. Series approximations seem to be more useful for a gradual contraction [B20,S21,S22] and Black and Denn [B21] have shown that, with an angle of taper of less than 45° and a contraction ratio of at least five, the streamlines for a Maxwell model are straight, so that sink flow is a reasonable approximation. The experiments of Clermont, Le Roy and Pierrard [C22] lend support to this conclusion for a smaller contraction ratio.

The importance of elongational flows in processing is emphasised by Denson [D17] and by Cogswell [C27] and, as we have seen, high stresses are likely to develop during elongation. Hence it is important to be able to make simple calculations if the elongational component of a flow is significant. Cogswell offers the simple rule that there is an elongational component if the streamlines are not parallel. This seems intuitively obvious and a formal proof might seem to show an excessive mathematical zeal, so we do not labour the point. Having decided that there is elongational flow, the calculation of stresses requires a model, and there are no commonly accepted simple formulae at present. The aim of determining some which are reliable and simple is to be encouraged, but we note that, as Takserman-Krozer [T3] observes on the basis of a flexible chain model of a macromolecule, "the introduction of shearing components into a velocity field with a parallel gradient leads not only to additional effects characteristic of the shearing field but also to the appearance of some interaction terms" which involve the product of the rate of shear strain and the rate of elongational strain. This is essentially a reminder that polymeric liquid behaviour is not linear, and any attempt at linearization should be carefully examined.

In addition to a number of papers at a recent symposium [C38], some of which we have cited separately above, the reader may find useful information in reviews by Forsyth [F12], Denn [D14] and Cogswell [C28]. Forsyth discusses a number of analyses of converging flows, without any detailed mention of the elongational flow aspect of these flows, while Denn gives a careful discussion of this aspect and points out inconsistencies which arise from the assumption of sink flow, notably that the observed stresses are much larger than is given by any prediction based on reasonable values of material parameters. Cogswell reviews experimental arrangements and methods of making relatively simple calculations from measured flow rates and pressure drops.

Given the present state of our understanding, it is not surprising that (in spite of the warning given above) Cogswell's technique is that of calculating separately the pressure drops arising because of shearing and elongation. There are different expressions for elongational strain rate, depending on whether the converging flow is taken to be sink flow or steady extension. Cogswell [C28] points out that the differences are not large for angles of convergence of 30° or less, but this may well merit more careful attention in view of the problem referred to above [D14]. The same problem is encountered by Hinch and Elata [H10] who have great difficulty in finding a constitutive equation to predict high enough stresses in a converging flow of dilute polymer solutions, given the assumption of sink flow.

As well as seeking new constitutive equations, as they suggest [H10], we should question the assumption of sink flow [D14]. The two flows illustrated in Figure 7.1 are both steady extensions (section 4.3.2) with the difference that pure shear (Figure 7.1(b)) is a steady rectilinear extension (section 4.3.1), with a constant strain rate, while sink flow (Figure 7.1(a)) has an increasing strain rate (section 4.2.2). The detailed calculation for a fair comparison between these flows has not been carried out. The differences in stresses which may arise with a Maxwell model have been discussed earlier (sections 2.2.3, 2.3.7, 4.2.2). In addition we note that in sink flow the streamlines generate surfaces on which there is no shear stress, which is not the case in uniform extension. Denn [D14] reports the opinion of Metzner that the shearing stresses will increase the tensile stress in this case, but again the calculation has not been done.

7.2 NEARLY EXTENSIONAL FLOWS

In search of more generality Metzner [M28] has given the name "extensional primary field (EPF) approximation" to the approximation in which off-diagonal terms in the rate of strain tensor are neglected, and discusses the ideas surrounding this idea and its possible application. This basic idea is still awaiting development and use, which Metzner envisages as giving the EPF approximation a status comparable to that of boundary layer theory or the lubrication approximation.

There are two recent papers which offer some hope of progress in the foundations of this study. Both Pountney and Walters [P31] and Huilgol [H18] adopt the idea (following the work of Pipkin and Owen on nearly viscometric flows) of postulating a deformation history which differs by a tensor $\underline{E}$ whose norm is small from the appropriate tensor for uniaxial elongation, e.g. (2.42b). The stress is then that for the elongational flow together with a perturbation linear in $\underline{E}$ with a possible 81 different components. This number is reduced using symmetry and frame indifference to three independent linear functionals.

These three functionals may be related to the elongational viscosity function, by using the fact that this function must be recovered if the perturbed flow is itself steady uniaxial elongation. Two relations are obtained by this technique between the functionals (evaluated for steady uniaxial elongation) and the elongational viscosity and its derivative with respect to the strain rate.

Since these results hold for any simple fluid, and for small deviations from elongational flow, they do not require the property which we have noted for many polymeric liquids that stresses in elongational flow are much greater than in shear flow. This property appears to underly the significance of elongation in many of the examples Metzner discusses in proposing the EPF approximation [M28]. Hence this author feels that, while the general theory of nearly extensional flows is important, there is a potentially more powerful idea to be uncovered in formalising and developing the EPF idea.

7.3 APPLICATIONS

We have already mentioned some of the obvious applications of elongational flows, in the manufacture of artificial fibres and films. The importance of the elongational behaviour of polymer melts in determining how they flow through converging passages has also been discussed above, and the reviews of stability [P5,P14] point to the probable significance of this in relation to "melt fracture" - a phenomenon of gross irregularities in extruded molten polymer.

Many moulding and forming processes involve substantial amounts of elongation [D17], for example vacuum forming involves the stretching of a softened sheet of polymer until it fits a mould, and the stretching may be quite different in different parts of the sheet. Blow moulding involves both the stretching of a tube of molten polymer under its own weight (while it is extruded in preparation for blowing) and the inflation of the tube into a mould, typically to produce a bottle. The elongation here, for a long bottle, is essentially two-dimensional (strip biaxial extension) and simplifications are possible in analysis of the process [P13]. Alfrey [A12] has discussed a number of processes of this type and indicates how analysis may proceed by modifying membrane theory (from solid mechanics) - see also work on film blowing [P4,P12,P6,H4,W1].

Squeezing flows, e.g. where liquid is contained between two parallel plates which approach one another, are another class of flows where shearing and elongation may both be significant. The discussion in section 1.1.1 gives formulae for the two extreme assumptions of shearing alone (the lubrication approximation) and elongation alone. It is interesting to discover that, according to White [W16] "The first problem one meets in rubber processing" involves the flow of a bale of rubber being squeezed by the weight of the bales stacked on top of it. This is analysed by White using lubrication theory, but data of Leider [L8] and of Brindley, Davies and Walters [B25] show that this does not work well initially, and we may suppose that the elongational part of the flow is relevant then. Williams and Tanner [W21] argued, on the basis of the network rupture model (2.37), that the elongational flow will not contribute significantly to the stresses, and that the squeeze film is not a suitable candidate for the EPF approximation [M28]. This seems to be the wrong way to argue, since they start from a model which does not predict particularly large stresses in elongation, contrary to evidence discussed above.

A manifestation of the elongational nature of converging flow is the alignment of fibres in a flowing suspension. This has important implications in, for example, the injection moulding of fibre-reinforced plastics and experimental studies have been undertaken to investigate the phenomenon [H6,L7]. The first of these studies includes a theoretical treatment based on suspension mechanics which is reasonably successful. This work has not been connected explicitly with work on elongational viscosity of suspensions [B8,M29,W8] which we have cited above (section 4.1.1). There is obviously a relation also to studies of elongational viscosity and of orientation of solutions of macromolecules, in particular a study of rod-like macromolecules [M14] which is relevant to recent developments in ultra-high-modulus fibres.

We can trace such work back to Ziabicki [Z6], Takserman-Krozer [T2,T3] and Peterlin [P11], and have discussed aspects of this above (section 4.1.1). Bird and co-workers have reviewed work in this area [B17,B19], and there is much current work concerned with modelling macromolecules in solution. Hinch [H9] reviews a number of ideas in developing a model which he seeks to use in explaining the phenomenon of drag reduction in turbulent flows due to small concentrations of dissolved polymer (the Toms phenomenon - see, for example, [L13]). This turns out to be associated with elongational flow, at least as far as the most widely accepted mechanism is concerned. This explanation is that the polymer solution has a greatly increased resistance to elongational flow compared to the solvent (water), and this serves to suppress some of the motion of turbulent eddies and thus either directly to reduce energy dissipation or (indirectly) to interfere with the energy transport from large to small eddies. A similarly large effect of a small amount of dissolved polymer has been observed in flow through porous media, and is similarly attributed to elongational flow, associated with the converging-diverging channels through porous media [J1,S7].

We can summarise this by observing that as far as dilute polymer solutions and suspensions of elongated particles are concerned, substantial differences in behaviour from that of the solvent are apparently always associated with elongational flow. Hence an appropriate rheological model must represent elongational behaviour well. The same is true for polymer melts and concentrated solutions, although the effects may not be as spectacular. There seems to be no doubt that both engineering and scientific interest in the study of elongational flows will continue to grow, and the mathematical problems arising from the analysis of elongational or nearly elongational flows offer many interesting challenges.

Bibliography

In the text we have cited a number of books; here we offer the reader a guide to a few which may be useful in extending his knowledge in a variety of directions. For basic continuum mechanics and rheology uncompromising mathematical treatments are offered by Lodge [L17] and Truesdell [T25] of which the former relates the theory to experimental results. Astarita and Marrucci [A17] offer an approach similar to Truesdell which is less formal, is more specifically concerned with liquid rather than solid behaviour, and, like Lodge, relates the theory to behaviour of real liquids. Probably the best account for the student or the theorist who lacks experience of the experimental evidence is that of Bird, Armstrong and Hassager [B18], whose second volume, by Bird, Armstrong, Hassager and Curtiss [B19] provides the only textbook offering an introduction to kinetic theory applied to the modelling of the rheological behaviour of polymeric liquids. Equally worthy of recommendation is Walters' book on rheometry [W3] and the two review articles of Pipkin and Tanner on viscometric flows [P27] and non-viscometric flows [P28], although the treatment of elongational flows in the latter feels somewhat dated, largely because of the rapid advances being made. Pipkin's treatment of viscoelasticity [P26] offers a useful mathematical treatment, with the flavour of elasticity and solid mechanics, while Ferry [F6] offers the classic presentation of linear theory and polymer viscoelasticity.

On polymer processing the only book devoted to elongational flow is by Ziabicki [Z9] on the manufacture of fibres. More general treatment with substantial coverage of processes involving elongations is given by Han [H5] who presents many relevant experimental results from his laboratory linked to theoretical treatments. Middleman [M32] considers the construction of the mathematical models in more detail in his latest book, which offers a contrasting treatment of many of the same processes that Han discusses. This is more likely to appeal to the mathematically minded, and represents much that this author feels should be taught to all engineering students. As well as his own opinion of these books, the author recommends several useful reviews which have appeared, including Metzner and Macosko on Walters [W3] in

AIChE J., 22 (1976), 414; Denn on Han [H5] in AIChE J., 23 (1977), 405; Cogswell on Han [H5] in J. Non-Newtonian Fluid Mech., 2 (1977), 397-398; Tanner on Astarita and Marrucci [A17] in J. Fluid Mech., 71 (1975), 207-208; and others in these journals or in Rheologica Acta and Rheology Abstracts.

On mathematical topics the selection is even more difficult and we mention only two out of many books which can introduce the reader to the qualitative analysis of nonlinear ordinary differential equations, by Hirsch and Smale [H11] and by Jordan and Smith [J9]. The text by Hale [H1] is one of many on more advanced aspects of this subject. A natural descendant of this line of work is catastrophe theory, and Poston and Stewart [P29] as well as providing an introduction to the application of this subject, reproduce much of the work of Berry and Mackley [B12] which we have discussed in the text.

Finally we mention the proceedings of five conferences, individual papers from which we have cited in the text. Theoretical Rheology [H19] is from the 1974 conference of the British Society of Rheology, which was joint organiser with the Plastics and Rubber Institute of Polymer Rheology and Plastics Processing [C20] in 1975. The Proceedings of the VIIth International Congress on Rheology [K15] contains brief accounts (generally two printed pages) of all the papers presented at Gothenburg in 1976, and on a smaller scale General Rheology and Stretching Flows [F5] does the same for the 1977 conference of the British Society of Rheology, part of which is published in full (Stretching Flows [P19]) as a special issue of the Journal of Non-Newtonian Fluid Mechanics. Non-Newtonian Fluid Mechanics is also the title of a 1978 symposium sponsored by IUTAM, the proceedings of which [C38] are to be published as volume 5 of that journal. The 49th Annual Meeting of the Society of Rheology (October 1978, Houston, Texas) promises at least six papers on aspects of elongational flows, evidence that activity proceeds apace and that this book would grow by several pages each month if it were allowed to do so.

Recollecting that the opening quotation of the Preface is addressed by *the Autocrat's* friend *the Professor* to *the Poet*, the author hopes he may be forgiven a final flight of fancy in suggesting the substitution of *rheology* for *poetry* in the following lines from *The Root of the Matter*.

> *There is poetry in everything. That*
> *is the biggest argument*
> *against poetry*
>
> *Miroslav Holub.*

References and author index

The references are listed alphabetically by first author's name and, where necessary, by date of publication. Citations in the text are enclosed in square brackets and include a letter (the initial letter of the first author's name) and a number, e.g. [L17]. The list of references here serves as an author index, with the pages on which references are cited listed at the end of the reference, e.g. *(pp.99,102)*. Authors other than the first-named are indexed by cross-reference to the detailed references in the same list.

L. E. Abbott [C14].

A1 D. Acierno, J. N. Dalton, J. M. Rodriguez and J. L. White, *Rheological and heat transfer aspects of the melt spinning of monofilament fibres of polyethylene and polystyrene.* J. Appl. Polym. Sci., 15 (1971), 2395-2415. *(pp.99,102)*

A2 D. Acierno, G. Titomanlio and R. Greco, *Differences in behaviour between dilute and concentrated polymer solutions in elongational flow.* Chem. Eng. Sci., 29 (1974), 1739-1744. *(pp.100, 101,104)*

A3 D. Acierno, G. Titomanlio and L. Nicodemo, *Elongational flow of dilute polymer solutions.* Rheol. Acta, 13 (1974), 532-537. *(pp.100,103,104)*

A4 D. Acierno, F. P. La Mantia, G. Marrucci and G. Titomanlio, *A nonlinear viscoelastic model with structure-dependent relaxation times. I. Basic formulation.* J. Non-Newtonian Fluid Mech., 1 (1976), 125-146. *(pp.29,31,49,50)*

A5 D. Acierno, F. P. La Mantia, G. Marrucci, G. Rizzo and G. Titomanlio, *A nonlinear viscoelastic model with structure-dependent relaxation times. II. Comparison with LDPE transient stress results.* J. Non-Newtonian Fluid Mech., 1 (1976), 147-157. *(pp.50,120,157)*

A6 D. Acierno, F. P. La Mantia and G. Titomanlio, *Model analysis of uniaxial and biaxial stretching of polymer melts.* Rheol. Acta, 15 (1976), 642-647. *(p.159)*

A7 D. Acierno, F. P. La Mantia and G. Marrucci, *A nonlinear viscoelastic model with structure-dependent relaxation times. III. Comparison with LD polyethylene creep and recovery data.* J. Non-Newtonian Fluid Mech., 2 (1977), 271-280. *(pp. 50,85,157)*

A8 D. Acierno, F. P. La Mantia, B. De Cindio and L. Nicodemo, *Transient shear and elongational data for polyisobutylene melts.* Trans. Soc. Rheol., 21 (1977), 261-271. *(p.160)*

A9 D. Acierno, G. Iorio, F. P. La Mantia and G. Marrucci, *Surface instabilities arising in drawing cylindrical specimens of LDPE melts.* J. Non-Newtonian Fluid Mech., 4 (1978), 99-109. *(p.157)*

A10 E. B. Adams and D. C. Bogue, *Viscoelasticity in shearing and accelerative flows: A simplified integral theory.* AIChE J., 16 (1970), 53-57. *(p.46)*

W. Aeschlimann [N9].

A11 P. K. Agrawal, W. K. Lee, J. M. Lorntson, C. I. Richardson, K. F. Wissbrun and A. B. Metzner, *Rheological behaviour of molten polymers in shearing and in extensional flows.* Trans. Soc. Rheol., 21 (1977), 355-379. *(pp.33,47,62, 72,73,74,76,80,81,83)*

V. M. Alekseyeva [V5].

A12 T. Alfrey Jr., *Fabrication of thermoplastic polymers.* Appl. Polym. Symp., 17 (1971), 3-24. *(p.210)*

T. Alfrey Jr. [W9].

A13 E. N. da C. Andrade, *On the viscous flow in metals and allied phenomena.* Proc. Roy. Soc., A84 (1910), 1-12. *(pp.5,6)*

A14 E. N. da C. Andrade and B. Chalmers, *The resistivity of polycrystalline wires in relation to plastic deformation, and the mechanism of plastic flow.* Proc. Roy. Soc., A138 (1932), 348-374. *(p.6)*

E. S. Andrews [T23].

B. E. Anshus [T20].

K. Aoki [I5].

R. C. Armstrong [B18,B19].

R. C. Ashton [O7].

A15 G. Astarita, *Two dimensionless groups relevant in the analysis of steady flows of viscoelastic materials.* Ind. Eng. Chem. Fundam. 6 (1967), 257-262. *(p.114)*

A16 G. Astarita and L. Nicodemo, *Extensional flow behaviour of polymer solutions.* Chem. Eng. J., 1 (1970), 57-66. *(pp.103,140)*

A17 G. Astarita and G. Marrucci, *Principles of Non-Newtonian Fluid Mechanics.* McGraw Hill, London (1974). *(pp.115,120,127,128,129,131, 213,214)*

A18 G. Astarita, *Is non-Newtonian fluid mechanics a culturally autonomous subject?* J. Non-Newtonian Fluid Mech., 1 (1976), 203-206. *(preface)*

G. Astarita [B4,M10,T20].

P. Avenas [D12].

B1 K. M. Baid and A. B. Metzner, *Rheological properties of dilute polymer solutions determined in extensional and in shearing experiments.* Trans. Soc. Rheol., 21 (1977), 237-260. *(pp.9,18,47,48,51,104)*

R. L. Ballman [E9,H21].

B2 E. D. Baily, *New extensional viscosity measurements on polyisobutylene.* Trans. Soc. Rheol., 18 (1974), 635-640. *(pp.72,84,88, 91)*

B3 R. L. Ballman, *Extensional flow of polystyrene melt.* Rheol. Acta, 4 (1965), 137-140. *(pp.72,82)*

B4 R. T. Balmer and G. Astarita, *Constant stretch history extensional flows.* Ing. Chim. Ital., 5 (1969), 101-107. *(pp.123,124,125, 134)*

B5 R. T. Balmer, *Steady and unsteady isochoric simple extensional flows.* J. Non-Newtonian Fluid Mech., 2 (1977), 307-322. *(pp.122,128)*

B6 R. T. Balmer and D. J. Hochschild, *Hyperbaric Fano flow of Newtonian fluids.* J. Rheol., 22 (1978), 165-180. *(p.96)*

B7 V. G. Bankar, J. E. Spruiell and J. L. White, *Melt spinning dynamics and rheological properties of nylon-6.* J. Appl. Polym. Sci., 21 (1977), 2135-2155. *(pp.95,98,102)*

B8 G. K. Batchelor, *The stress generated in a non-dilute suspension of elongational particles by pure straining motions.* J. Fluid Mech., 46 (1971), 813-829. *(pp.106,211)*

B9 R. K. Bayer, H. Schreiner and W. Ruland, *Tensile properties of an extruded polyethylene melt.* Rheol. Acta, 17 (1978), 28-32. *(p.102)*

J. Beautemps [V13].

J. C. Bellinger [K4,K5].

B10 A. Bergonzoni and A. J. Di Cresce, *The phenomenon of draw resonance in polymeric melts. Part 1. Qualitative view. Part 2. Correlation to molecular parameters.* Polym. Eng. Sci., 6 (1966), 45-49, 50-59. *(p.103)*

B11 B. Bernstein, E. A. Kearsley and L. Zapas, *A study of stress relaxation with finite strain.* Trans. Soc. Rheol., 7 (1963), 391-410. *(p.34)*

B. Bernstein [G13].

B12 M. V. Berry and M. R. Mackley, *The six roll mill; unfolding an unstable persistently extensional flow.* Phil. Trans. Roy. Soc., A287 (1977), 1-16. *(pp.56,116,117,214)*

M. Bezy [C39].

B13 M. Biermann, *Foundations of extensional viscometry, Part I, Prolegomena on motion groups encompassing viscometric motions.* Acta Mech., 11 (1971), 283-298. *(p.118)*

B14 G. Binder and F. H. Muller, *Nonlinear deformation. I. Compression of cylindrical specimens.* Koll. Z., 177 (1961), 129-149. *(p.7)*

G. Binder [M35].

B15 E. C. Bingham, *Plasticity and elasticity.* J. Franklin Inst., 197 (1924), 99-115. *(p.21)*

B16 R. B. Bird and P. Carreau, *A nonlinear viscoelastic model for polymer solutions and melts - I.* Chem. Eng. Sci., 23 (1968), 427-434. *(pp.31,48)*

B17 R. B. Bird, M. W. Johnson Jr. and J. F. Stevenson, *Molecular theories of elongational viscosity.* Proc. 5th Int. Cong. Rheol., 4 (1970), 159-168. *(p.211)*

B18 R. B. Bird, R. C. Armstrong and O. Hassager, *Dynamics of polymeric liquids, Vol. 1. Fluid mechanics.* Wiley (1977). *(pp.35,101,120,213)*

B19 R. B. Bird, O. Hassager, R. C. Armstrong and C. F. Curtiss, *Dynamics of polymeric liquids. Vol. II. Kinetic theory.* Wiley (1977). *(pp.28,111,211,213)*

R. B. Bird [S16].

B20 J. R. Black, M. M. Denn and G. C. Hsiao, *Creeping flow of a viscoelastic liquid through a contraction: a numerical perturbation solution.* Theoretical Rheology, [H19] (1975), 3-30. *(p.207)*

B21 J. R. Black and M. M. Denn, *Converging flow of a viscoelastic liquid.* J. Non-Newtonian Fluid Mech., 1 (1976), 83-92. *(p.207)*

R. Bloch [C10].

B22 L. L. Blyler and C. Gieniewski, *Experimental studies on fibre spinning of a polymer melt.* 47th Annual Meeting, Society of Rheology, New York (1977), Abstract C6. *(p.102)*

D. V. Boger [N3].

D. C. Bogue [A10,C13,C14,M18,T1].

B23 D. B. Bogy, *Use of one-dimensional Cosserat theory to study instability in a viscous liquid jet.* Phys. Fluids, 21 (1978), 190-197. *(pp.19,199)*

C. Bonnebat [V12,V13].

W. J. Bontinck [D20].

B24 J. Boow and W. E. S. Turner, *The viscosity and working characteristics of glasses. I.* J. Soc. Glass Technol. Trans., 26 (1942), 215-240. *(p.5)*

R. Bragg [O4].

B25 G. Brindley, J. M. Davies and K. Walters, *Elasticoviscous squeeze films. Part I.* J. Non-Newtonian Fluid Mech., 1 (1976), 19-37. *(p.211)*

B26 J. M. Broadbent, *Elastic liquid experiments with the four-roll mill and the two-roll mill.* VIIth Int. Cong. Rheol., [K15] (1976), 256-257. *(p.56)*

B27 F. Bueche, *The viscoelastic properties of plastics.* J. Chem. Phys., 22 (1954), 603-609. *(p.7)*

B28 F. Bueche, *Viscoelasticity of polymethylmethacrylates.* J. Appl. Phys., 26 (1955), 738-749. *(pp.7,85)*

B29 J. M. Burgers, in *First Report on Viscosity and Plasticity, Ch. II, Remarks in connection with the experimental investigation of flow properties.* Academy of Sciences, Amsterdam (1935), 73-109. *(p.8)*

B30 W. F. Busse, *Mechanical structures in polymer melts. I. Measurements of melt strength and elasticity. II. Roles of entanglements in viscosity and elastic turbulence.* J. Polym. Sci., A2,5 (1967), 1249-1259, 1261-1281. *(pp.97,103)*

J. F. Carley [S3].

C1 W. H. Carothers and J. W. Hill, *Studies of polymerization and ring formation. XV Artificial fibers from synthetic linear condensation super-polymers.* J. Amer. Chem. Sco., 54 (1932), 1579-1587. *(p.11)*

C2 W. H. Carothers and F. J. Van Natta, *Studies of polymerization and ring formation. XVIII. Polyesters from ω-Hydroxydecanoic acid.* J. Amer. Chem. Soc., 55 (1933), 4714-4719. *(p.12)*

C3 P. J. Carreau, *Rheological equations from molecular network theories.* Trans. Soc. Rheol., 16 (1972), 99-127. *(pp.32,46)*

P. J. Carreau [B16].

M. M. Carroll [C15].

J. M. Carruthers [C28].

B. Caswell [N4].

C4 E. Cernia and W. Conti, *On the dynamics of melt spinning.* J. Appl. Polym. Sci., 17 (1973), 3637-3650. *(pp.98,102)*

B. Chalmers [A14].

N. H. Chamberlain [C36].

C. F. Chan Man Fong [M26].

C5 Y. Chan, J. L. White and Y. Oyanagi, *A fundamental study of the rheological properties of glass fiber reinforced polyethylene and polystyrene melts.* J. Rheol., 22 (1978), 507-524. *(pp.83,84,87)*

C6 S. Chandrasekhar, *Hydrodynamic and hydromagnetic stability.* Oxford University Press (1961). *(p.18)*

C7 H. Chang and A. S. Lodge, *A possible mechanism for stabilising elongational flow in certain polymeric liquids at constant temperature and composition.* Rheol. Acta, 10 (1971),

C7 H. Chang and A. S. Lodge - *continued*
448-449. *(pp.18,159)*

C8 H. Chang and A. S. Lodge, *Comparison of rubberlike-liquid theory with stress-growth data for elongation of a low-density branched polyethylene melt.* Rheol. Acta 11 (1972), 127-129. *(pp.79,81,155,156)*

C9 J.-C. Chang and M. M. Denn, *Continuous drawing of Newtonian and visco-elastic liquids.* IUTAM preprints, [C38] (1978). *(pp.104,202,205)*

C10 W. V. Chang, R. Bloch and N. W. Tschoegl, *On the theory of the visco-elastic behaviour of soft polymers in moderately large deformations.* Rheol. Acta, 15 (1976), 367-378. *(p.30)*

C11 J. M. Charrier and Y. K. Li, *Large elastic deformation theories and unstable deformation of rubber structures: thick tubes.* Trans. Soc. Rheol., 21 (1977), 301-325. *(p.92)*

C12 A. Chaure, *On motions with constant stretch history.* C. R. Acad. Sci. Paris, A274 (1972), 1839-1842. *(p.133)*

C13 I.-J. Chen and D. C. Bogue, *Time-dependent stress in polymer melts and review of viscoelastic theory.* Trans. Soc. Rheol., 16 (1972), 59-78. *(pp.31,33,48)*

C14 I.-J. Chen, G. E. Hagler, L. E. Abbott, D. C. Bogue and J. L. White, *Interpretation of tensile and melt spinning experiments on low density and high density polyethylene.* Trans. Soc. Rheol., 16 (1972), 473-494. *(pp.83,84,98,99,102)*

C15 S. C. Chow and M. M. Carroll, *Motions of proportional extension.* Quart.J. Mech. Appl. Math., 26 (1973), 471-482. *(p.126)*

C16 S. C.-K. Chung and J. F. Stevenson, *A general elongational flow experiment: inflation and extension of a viscoelastic tube.* Rheol. Acta, 14 (1975), 832-841. *(p.124)*

S. C.-K. Chung [S19].

A. Ciferri [M14].

C17 B. de Cindio, L. Nicodemo, L. Nicolais and S. Ranaudo, *Elongational viscosity of suspensions.* VIIth Int. Cong. Rheol., [K15] (1976), 204-205. *(p.103)*

B. de Cindio [A8,N5].

C18 N. S. Clarke, *Two-dimensional flow under gravity in a jet of viscous liquid.* J. Fluid Mech., 31 (1968), 481-500. *(p.197)*

C19 N. S. Clarke, *The asymptotic effects of surface tension and viscosity on an axially-symmetric free jet of liquid under gravity.* Quart. J. Mech. Appl. Math., 22 (1969), 247-256. *(pp.192,196)*

C20 P. L. Clegg, F. N. Cogswell, D. E. Marshall and S. G. Maskell, editors, *Polymer Rheology and Plastics Processing - Conference Proceedings.* Plastics and Rubber Institute, London (1975). *(p.214)*

C21 J. R. Clermont, J. M. Pierrard and O. Scrivener, *Experimental study of a non-viscometric flow: kinematics of a viscoelastic fluid at the exit of a cylindrical tube*. J. Non-Newtonian Fluid Mech., 1 (1976), 175-182. *(p.189)*

C22 J. R. Clermont, P. Le Roy and J. M. Pierrard, *Theoretical and experimental study of a viscoelastic fluid in the converging region of a pipe*. IUTAM preprints [C38] (1978). *(pp.9,208)*

C23 A. F. Clift, *Observations on certain rheological properties of human cervical secretion*. Proc. Roy. Soc. Med., 39 (1945), 1-9. *(p.11)*

C24 F. N. Cogswell, *The rheology of polymer melts under tension*. Plast. Polym., 36 (1968), 109-111. *(pp.5,72,83)*

C25 F. N. Cogswell, *Tensile deformations in molten polymers*. Rheol. Acta 8 (1969), 187-194. *(pp.72,83,85,103)*

C26 F. N. Cogswell and D. R. Moore, *A comparison between simple shear, elongation and equal biaxial extension deformations*. Polym. Eng. Sci., 14 (1974), 573-576. *(pp.85,91)*

C27 F. N. Cogswell, *Polymer melt rheology during elongational flow*. Appl. Polym. Symp., 27 (1975), 1-18. *(pp.76,208)*

C28 F. N. Cogswell, *Converging flow and stretching flow: a compilation*. J. Non-Newtonian Fluid Mech., 4 (1978), 23-38. *(p.208)*

F. N. Cogswell [C20].

C29 R. E. Cohen and J. M. Carruthers, *Stress and temperature dependence of entanglement slippage in cis-1,4 Polybutadiene*. 46th Annual Meeting, Society of Rheology, St. Louis, Missouri (1975), Abstract 6A. *(p.87)*

C30 B. D. Coleman and W. Noll, *Recent results in the continuum theory of viscoelastic fluids*. Ann. N.Y. Acad. Sci., 89 (1961), 672-714. *(pp.35,120)*

C31 B. D. Coleman and W. Noll, *Steady extension of incompressible simple fluids*. Phys. Fluids, 5 (1962), 840-843. *(pp.118,119, 121,135)*

C32 B. D. Coleman, *Kinematical concepts with applications in the mechanics and thermodynamics of incompressible viscoelastic fluids*. Arch. Rat. Mech. Anal., 9 (1962), 273-300. *(p.134)*

C33 B. D. Coleman, *Substantially stagnant motions*. Trans. Soc. Rheol., 6 (1962), 293-300. *(p.134)*

C34 B. D. Coleman, *On the use of symmetry to simplify the constitutive equations of isotropic materials with memory*. Proc. Roy. Soc., A306 (1968), 449-476. *(pp.122,123,125,127, 128,129,131)*

C35 B. D. Coleman and E. H. Dill, *On the stability of certain motions of incompressible materials with memory*. Arch. Rat. Mech. Anal., 30 (1968), 197-224. *(pp.125,126)*

R. W. Connelly [G1].

W. Conti [C4].

C36 M. Copley and N. H. Chamberlain, *Filament attenuation in melt spinning and its effect on axial temperature gradient.* Appl. Polym. Symp., 6 (1967), 27-50. *(p.103)*

F. M. V. Coppen [S4].

D. L. Crady [D18].

T. Craft [Z3].

A. J. di Cresce [B10].

C37 M. J. Crochet and G. Pilate, *Plane flow of a fluid of second grade through a contraction.* J. Non-Newtonian Fluid Mech., 1 (1976), 247-258. *(p.207)*

C38 M. J. Crochet, editor, *Non-Newtonian Fluid Mechanics. Preprints of a IUTAM symposium, Louvain-la-Neuve, 28 August-1 September 1978.* (To appear as J. Non-Newtonian Fluid Mech., 5 (1979).) *(pp.208,214)*

C39 M. J. Crochet and M. Bezy, *Numerical solutions for the flow of visco-elastic fluids.* IUTAM preprints [C38] (1978). *(p.207)*

J. A. Cuculo [H7].

C. F. Curtiss [B19].

D1 C. A. Dahlquist and M. R. Hatfield, *Constant stress elongation of soft polymers: time and temperature studies.* J. Coll. Sci., 7 (1952), 253-267. *(pp.6,84,86)*

J. N. Dalton [A1].

D2 J. M. Davies, J. F. Hutton and K. Walters, *A critical re-appraisal of the jet-thrust technique, with particular reference to axial velocity and stress rearrangement at the exit plane.* J. Non-Newtonian Fluid Mech., 3 (1977), 141-160. *(p.189)*

J. M. Davies [B25].

D3 J. M. Dealy, *Extensional flow of non-Newtonian fluids - a review.* Polym. Eng. Sci., 11 (1971), 433-445. *(pp.46,120)*

D4 J. M. Dealy, *On the relationship between extensional viscosities for uniaxial and biaxial extension.* Trans. Soc. Rheol., 17 (1973), 255-258. *(p.37)*

D5 J. M. Dealy, *New methods for measuring rheological properties of molten polymers.* Polym. Rheol. & Plastics Processing, [C20] (1975), 35-48. *(pp.88,89)*

D6 J. M. Dealy, R. Farber, J. Rhi-Sausi and L. Utracki, *Experience with a constant stress melt extensiometer.* Trans. Soc. Rheol., 20 (1976), 455-464. *(pp.5,71,72)*

D7 J. M. Dealy, *Extensional rheometers for molten polymers: a review.* J. Non-Newtonian Fluid Mech., 4 (1978), 9-21. *(pp.71,72)*

D8 J. M. Dealy and J. Rhi-Sausi, *Biaxial stretching of molten polyethylene.* For presentation at the AIChE meeting, Miami Beach, November 1978. *(pp.88,89,92)*

J. M. Dealy [R6].

D9 J. R. Dees and J. E. Spruiell, *Structure development during melt spinning of linear polyethylene fibers.* J. Appl. Polym. Sci., 18 (1974), 1053-1078. *(pp.97,103)*

D10 M. M. Denn and G. Marrucci, *Stretching of viscoelastic liquids.* AIChE J., 17 (1971), 101-103. *(pp.44,152)*

D11 M. M. Denn, *Stability of Reaction and Transport Processes.* Prentice-Hall, Englewood Cliffs, N.J. (1975). *(p.205)*

D12 M. M. Denn, C. J. S. Petrie and P. Avenas, *Mechanics of steady spinning of a viscoelastic liquid.* AIChE J., 21 (1975), 791-799. *(pp.67,164,165,166,172,177,180,184)*

D13 M. M. Denn and G. Marrucci, *Effect of a relaxation time spectrum and mechanics of polymer melt spinning.* J. Non-Newtonian Fluid Mech., 2 (1977), 159-168. *(pp.155,177,178,179,183)*

D14 M. M. Denn, *Extensional flows: experiment and theory*, in *The Mechanics of Viscoelastic Fluids* ed. R. S. Rivlin, AMD-Volume 22, ASME, New York (1977), 101-125. *(pp.111,155,208,209)*

D15 M. M. Denn, *On an analysis of draw resonance by Hyun.* AIChE J., (submitted, September 1978). *(p.202)*

M. M. Denn [B20,B21,C9,F8,F9,F10,P14].

D16 C. D. Denson and R. J. Gallo, *Measurements on the biaxial extension viscosity of bulk polymers: the inflation of a thin polymer sheet.* Polym. Eng. Sci., 11 (1971), 174-176. *(pp.88,91,159)*

D17 C. D. Denson, *Implications of extensional flow in polymer fabrication processes.* Polym. Eng. Sci., 13 (1973), 125-130. *(pp.208,210)*

D18 C. D. Denson and D. L. Crady, *Measurements on the planar extensional viscosity of bulk polymers: The inflation of a thin rectangular polymer sheet.* J. Appl. Polym. Sci., 18 (1974), 1611-1618. *(pp.88,89,91)*

D19 C. D. Denson and D. C. Hylton, *A rheometer for measuring the viscoelastic response of polymer melts in planar and biaxial extensional deformations.* VIIth Int. Cong. Rheol., [K15] (1976), 386-387. *(pp.88,89,92)*

C. D. Denson [J12,J13].

D20 E. Deprez and W. J. Bontinck, *Rheological behaviour of polyethylene melts in elongational flow.* Polym. Rheol. & Plastics Processing, [C20] (1975), 274-287. *(pp.98,101,102,103)*

D21 R. A. Dickie and T. L. Smith, *Viscoelastic properties of a rubber vulcanizate under large deformations in equal biaxial tension, pure shear, and simple tension.* Trans. Soc. Rheol., 15 (1971), 91-110. *(p.92)*

R. A. Dickie [S13].

E. H. Dill [C35].

D22 L. K. Djiauw and A. N. Gent, *Elasticity of uncrosslinked rubbers*. J. Polym. Sci., Symp., 48 (1974), 159-168. *(pp.85,86,87)*

D23 M. Doi and S. F. Edwards, *Dynamics of concentrated polymer systems. Part 4. Rheological properties*. J. Chem. Soc. Faraday Trans. II, (in press, November 1978). (See also *Parts 1,2,3,* ibid., 74 (1978) 1789-1832.) *(p.47)*

D24 G. J. Donnelly and C. B. Weinberger, *Stability of isothermal spinning of a Newtonian fluid*. Ind. Eng. Chem. Fundam., 14 (1975), 334-337. *(pp.202,205)*

S. F. Edwards [D23].

C. Elata [H10].

E1 E. J. Elgood, N. S. Heath, B. J. O'Brien and D. H. Solomon, *Effect of molecular weight distribution on the sprayability of polymethyl methacrylate*. J. Appl. Polym. Sci., 8 (1964), 881-887. *(p.19)*

C. Engelter [M36].

E2 S. English, *The effect of various constituents on the viscosity of glass near its annealling temperature*. J. Soc. Glass Technol. Trans., 7 (1923), 25-45. *(p.4)*

E3 S. English, *The effect of composition on the viscosity of glass, Part II*. J. Soc. Glass Technol. Trans., 8 (1924), 205-248. *(p.4)*

E4 S. English, *The viscosity of glass*. J. Soc. Glass Technol., 12 (1928), 106-113. *(p.5)*

E5 H. Erbring, *Investigations of the spinnability of fluid systems*, Koll.-Beih., 44 (1936), 171-237. *(pp.10,13)*

E6 H. Erbring, *Information on the spinnability of lyophilic colloidal solutions and mechanical properties of solid fibres prepared from them. II*. Koll.-Z., 77 (1936), 32-36. *(pp.10,12)*

E7 H. Erbring, *On the spinnability of liquids (Part III)*. Koll.-Z., 77 (1936), 213-219. *(p.10)*

E8 A. E. Everage Jr. and R. J. Gordon, *On the stretching of dilute polymer solutions*. AIChE J., 17 (1971), 1257-1259. *(p.152)*

E9 A. E. Everage Jr. and R. L. Ballman, *Extensional viscosity of amorphous polystyrene*. J. Appl. Polym. Sci., 20 (1976), 1137-1141. *(pp.74,83)*

A. E. Everage Jr. [G16].

F1 G. Fano, *Contribution to the study of thread-forming materials*. Arch. fisiol., 5 (1908), 365-370. *(pp.9,11,96)*

R. Farber [D6].

F2 G. M. Fehn, *Steady-state drawing of polymer melts*. J. Polym. Sci. A-1, 6 (1968), 247-251. *(pp.95,99,102)*

F3 G. M. Fehn, *An empirical model for PVC melt drawing*. Rheol. Acta, 13 (1974), 767-773. *(p.102)*

F4 J. Ferguson and N. E. Hudson, *A new viscometer for the measurement of apparent elongational viscosity*. J. Phys. E, 8 (1975), 526-530. *(pp.95,99)*

F5 J. Ferguson, N. E. Hudson and C. J. S. Petrie, editors, *Summarised Proceedings of the British Society of Rheology Conference on General Rheology and Stretching Flows, Edinburgh, 7-9 September 1977*. British Society of Rheology, c/o Department of Fibre Science, University of Strathclyde, Glasgow. *(p.214)*

J. Ferguson [H13,H14,M5].

F6 J. D. Ferry, *Viscoelastic properties of polymers*, 2nd edition. Wiley, New York (1970). *(pp.142,213)*

J. D. Ferry [G19,K17,N11,T14].

V. D. Fikhman [R1,V2,V3,V4,V5].

F7 W. N. Findley and J. S. Y. Lai, *A modified superposition principle applied to creep of nonlinear viscoelastic material under abrupt changes in state of combined stress*. Trans. Soc. Rheol., 11 (1967), 361-380. *(p.86)*

F8 R. J. Fisher and M. M. Denn, *Finite-amplitude stability and draw resonance in isothermal melt spinning*. Chem. Eng. Sci., 30 (1975), 1129-1134. *(p.202)*

F9 R. J. Fisher and M. M. Denn, *A theory of isothermal melt spinning and draw resonance*. AIChE J., 22 (1976), 236-246. *(pp.177, 179)*

F10 R. J. Fisher and M. M. Denn, *Mechanics of nonisothermal polymer melt spinning*. AIChE J., 23 (1977), 23-28. *(pp.182,205)*

F11 R. J. Flowers and G. Lianis, *Viscoelastic tests under finite strain and variable strain rate*. Trans. Soc. Rheol., 14 (1970), 441-459. *(pp.87,92)*

S. J. Folley [S4].

F12 T. H. Forsyth, *Converging flow of polymers*. Polym.-Plast. Technol. Eng., 6 (1976), 101-131. *(p.208)*

F13 F. C. Frank, *The strength and stiffness of polymers*. Proc. Roy. Soc., A319 (1970), 127-136. *(pp.106,111)*

J. E. Frederick [S12].

A. Freudenthal [R2].

T. Fujimura [I7].

R. J. Gallo [D16].

G1 L. J. Garfield, G. H. Pearson and R. W. Connelly, *Analysis of a fixed-end constant-length melt elongation experiment*. 47th

G1 L. J. Garfield, G. H. Pearson and R. W. Connelly - *continued*
Annual Meeting, Society of Rheology, New York (1977), Abstract E15. *(pp.74,83)*

G2 F. H. Garner, A. H. Nissan and G. F. Wood, *Thermodynamics and rheological behaviour of elastoviscous systems under stress.* Phil. Trans. Roy. Soc., A243 (1950), 37-66. *(p.13)*

G3 F. H. Gaskins and W. Philippoff, *The behaviour of jets of viscoelastic fluids.* Trans. Soc. Rheol., 3 (1959), 181-203. *(p.13)*

J. Gavis [G11].

G4 P. G. de Gennes, *Coil-stretch transition of dilute flexible polymers under ultrahigh velocity gradients.* J. Chem. Phys., 60 (1974), 5030-5042. *(pp.106,108,109,110,111,114)*

A. N. Gent [D22].

H. H. George [L7].

G5 F. T. Geyling, *Basic fluid dynamic considerations in drawing optical fibers.* Bell System Tech., J., 55 (1976), 1011-1056. *(pp.17,18)*

C. Gienewski [B22].

G6 H. Giesekus, *Flows with constant velocity gradients and the movement of suspended particles in them. Part I. Three dimensional flows.* Rheol. Acta, 2 (1962), 101-112. *(pp.105,107, 108,112)*

G7 H. Giesekus, *Flows with constant velocity gradients and the movement of suspended particles in them. Part II. Plane flows and an experimental arrangement for their realisation.* Rheol. Acta, 2 (1962), 112-122. *(pp.56,112,114)*

G8 H. Giesekus, *Fluids with singular flow properties at rest (quasi-plastic fluids).* Rheol. Acta, 2 (1962), 122-130. *(p.105)*

G9 H. Giesekus, *Statistical rheology of suspensions and solutions with special reference to normal stress effects* in *Second Order Effects in Elasticity, Plasticity and Fluid Dynamics* , ed. M. Reiner and D. Abir, Pergamon, Oxford (1964), 553-584. *(pp.56,114)*

G10 H. Giesekus, *A symmetric formulation of the linear theory of viscoelastic materials.* Proc. 4th Int. Cong. Rheol., 3 (1965), 15-28. *(pp.23,77,142)*

G11 S. J. Gill and J. Gavis, *Tensile stresses in jets of viscoelastic fluids. I.* J. Polym. Sci., 20 (1956), 287-296. *(p.13)*

G12 J. D. Goddard, *A modified functional expansion for viscoelastic fluids.* Trans. Soc. Rheol., 11 (1967), 381-399. *(p.35)*

J. D. Goddard [W8].

G13 W. Goldberg, B. Bernstein and G. Lianis, *The exponential extension rate history, comparison of theory with experiment.* Int. J. Nonlinear Mech., 4 (1969), 277-300. *(pp.84,87)*

G14 M. Goldin, J. Yerushalmi, R. Pfeffer and R. Shinnar, *Breakup of a laminar capillary jet of a viscoelastic fluid.* J. Fluid Mech., 38 (1969), 689-711. *(p.18)*

G15 M. Gordon, J. Yerushalmi and R. Shinnar, *Instability of jets of non-Newtonian fluids.* Trans. Soc. Rheol., 17 (1973), 303-324. *(p.18)*

G16 R. Gordon and A. E. Everage Jr., *Bead-spring model of dilute polymer solutions: continuum modifications and an explicit constitutive equation.* J. Appl. Polym. Sci., 15 (1971), 1903-1909. *(p.28)*

R. J. Gordon [E8].

G17 V. A. Gorodtsov and A. I. Leonov, *On the kinematics, nonequilibrium thermodynamics and rheological relationships in the nonlinear theory of viscoelasticity.* J. Appl. Math. Mech., 32 (1968), 62-84. *(p.7)*

G18 D. D. Goulden and W. C. MacSporran, *An experimental study of nonrheometric flows of viscoelastic fluids. II. Flow in a liquid jet exiting from a vertical tube.* J. Non-Newtonian Fluid Mech., 1 (1976), 183-198. *(p.189)*

G19 R. Greco, C. R. Taylor, O. Kramer and J. D. Ferry, *Rubber networks containing unattached macromolecules. III. Nonlinear stress relaxation in the system Ethylene-Propylene terpolymer with ethylene-propylene copolymer and behaviour of extracted networks.* J. Polym. Sci.: Polym. Phys., 13 (1975), 1687-1694. *(p.86)*

R. Greco [A2,K17,T14].

G20 A. E. Green and R. S. Rivlin, *The mechanics of nonlinear materials with memory.* Arch. Rat. Mech. Anal., 1 (1957), 1-21. *(p.35)*

G21 A. E. Green and N. Laws, *Ideal fluid jets.* Int. J. Eng. Sci., 6 (1968), 317-328. *(p.199)*

G22 A. E. Green, *On the nonlinear behaviour of fluid jets.* Int. J. Eng. Sci., 14 (1976), 49-63. *(p.199)*

D. R. Gregory [W4].

G23 G. S. Gunter, *The determination of "spinnbarkeit" of synovial fluid and its destruction by enzymic action.* Austral. J. Expt. Biol. Med. Sci., 27 (1949), 256-274. *(p.10)*

G24 P. A. Gutteridge, *Stretching flows for thin film production. 2. Film stretching over a conical mandrel.* J. Non-Newtonian Fluid Mech., 4 (1978), 73-88. *(p.159)*

P. A. Gutteridge [P6].

G. E. Hagler [C14].

H1 J. K. Hale, *Ordinary differential equations.* Wiley-Interscience, New York (1969). *(pp.171,214)*

H2 C. D. Han and R. R. Lamonte, *Studies on melt spinning. I. Effect of molecular structure and molecular weight distribution on elongational viscosity.* Trans. Soc. Rheol., 16 (1972), 447-472. *(pp.101,102)*

H3 C. D. Han and Y. W. Kim, *Studies on melt spinning. V. Elongational viscosity and spinnability of two-phase systems.* J. Appl. Polym. Sci., 18 (1974), 2589-2603. *(p.102)*

H4 C. D. Han and J. Y. Park, *Studies on blown film extension. I. Experimental determination of elongational viscosity.* J. Appl. Polym. Sci., 19 (1975), 3257-3276. *(pp.90,91,159, 210)*

H5 C. D. Han, *Rheology in polymer processing.* Academic Press, New York (1976). *(pp.213,214)*

R. J. Hansen [L13].

H6 J. B. Harris and J. F. T. Pittman, *Alignment of slender rod-like particles in suspensions using converging flow.* Trans. Inst. Chem. Eng., 54 (1976), 73-83. *(p.211)*

O. Hassager [B18,B19].

M. R. Hatfield [D1].

J. Hauenstein [L12].

N. S. Heath [E1].

C. T. Hill [N2].

H7 J. W. Hill and J. A. Cuculo, *An experimental study of threadline dynamics with emphasis on the effect of molecular weight on the elongational viscosity of melt-spun polyethylene terephalate.* J. Appl. Polym. Sci., 18 (1974), 2569-2588. *(pp.98,102)*

J. W. Hill [C1].

H8 E. J. Hinch, *Mechanical models of dilute polymer solutions for strong flows with large polymer deformations.* Colloques Int. du C.N.R.S., 233 (1975), 241-247. *(pp.106,108,111)*

H9 E. J. Hinch, *Mechanical models of dilute polymer solutions in strong flows.* Phys. Fluids, 20 (1977), S22-S30. *(p.211)*

H10 E. J. Hinch and C. Elata, *Heterogeneity of dilute polymer solutions.* IUTAM preprints, [C38] (1978). *(p.209)*

T. Hiraoka [K14].

H11 M. W. Hirsch and S. Smale, *Differential equations, dynamical systems and linear algebra.* Academic Press, New York (1974). *(p.214)*

D. J. Hochschild [B6].

D. L. Holt [L1,L2].

H12 K. C. Hoover and R. W. Tock, *Experimental studies on the biaxial extensional viscosity of polypropylene.* Polym. Eng. Sci., 16 (1976), 82-86. *(p.92)*

M. Horie [V11].

C. B. Howard [K1].

G. C. Hsiao [B20].

H13 N. E. Hudson, J. Ferguson and P. Mackie, *The measurement of the elongational viscosity of polymer solutions using a viscometer of novel design.* Trans. Soc. Rheol., 18 (1974), 541-562. *(pp.99,104)*

H14 N. E. Hudson and J. Ferguson, *Correlation and molecular interpretation of data obtained in elongational flow.* Trans. Soc. Rheol., 20 (1976), 265-286. *(pp.99,104)*

N. E. Hudson [F4,F5,M5].

H15 R. R. Huilgo., *On a characterization of simple extensional flows.* Rheol. Acta, 14 (1975), 48-50. *(p.135)*

H16 R. R. Huilgol, *Algorithms for motions with constant stretch history.* Rheol. Acta, 15 (1976), 120-129. *(p.135)*

H17 R. R. Huilgol, *Algorithms for motions with constant stretch history - II.* Rheol. Acta, 15 (1976), 577-578. (See also Rheol. Acta, 17 (1978), 460.) *(p.135)*

H18 R. R. Huilgol, *Nearly extensional flows.* IUTAM preprints, [C38] (1978). *(p.209)*

R. R. Huilgol [T12].

D. L. Hunston [L13,T19].

J. D. Huppler [S16].

H19 J. F. Hutton, J. R. A. Pearson and K. Walters, editors, *Theoretical rheology.* Applied Science Publishers, Barking, Essex (1975). *(p.214)*

J. F. Hutton [D2].

D. C. Hylton [D19].

H20 J. C. Hyun, *Theory of draw resonance. Part I. Newtonian fluids. Part II. Power-law and Maxwell fluids.* AIChE J., 24 (1978), 418-426. *(p.202)*

H21 J. C. Hyun and R. L. Ballman, *Isothermal melt spinning - Lagrangian and Eulerian viewpoints.* J. Rheol., 22 (1978), 349-380. *(pp.62,164)*

K. S. Hyun [K5].

I1 IBM J. Res. Develop., 21 2-80, *Special issue devoted to ink-jet printing.* *(p.19)*

I2 Y. Ide and J. L. White, *The spinnability of polymer fluid filaments.* J. Appl. Polym. Sci., 20 (1976), 2511-2531. *(pp.16,74,205)*

I3 Y. Ide and J. L. White, *Investigation of failure during elongational flow of polymer melts.* J. Non-Newtonian Fluid Mech., 2 (1977), 281-298. *(pp.16,74,205)*

I4 Y. Ide and J. L. White, *Experimental study of elongational flow and failure of polymer melts.* J. Appl. Polym. Sci., 22 (1978), 1061-1079. *(pp.74,75,83,84,85)*

Y. Ide [W13,W18].

G. Iorio [A9].

I5 T. Ishibashi, K. Aoki and T. Ishii, *Studies on melt spinning of Nylon-6. I. Cooling and deformation behaviour and orientation of Nylon-6 threadline.* J. Appl. Polym. Sci., 14 (1970), 1597-1613. *(pp.97,102)*

I6 H. Ishihara and S. Kase, *Studies on melt spinning, V. Draw resonance as a limit cycle.* J. Appl. Polym. Sci., 19 (1975), 557-566. *(pp.202,205)*

T. Ishii [I5].

I7 K. Iwakura, M. Yoshinari and T. Fujimura, *A method for the measurement of the planar extensional viscosity.* J. Soc. Rheol. Japan, 3 (1975), 134-136. *(pp.90,92)*

J1 D. F. James and D. R. McLaren, *The laminar flow of dilute polymer solutions through porous media.* J. Fluid Mech., 70 (1975), 733-752. *(p.212)*

J2 A. Jeffrey, *Quasilinear hyperbolic systems and waves.* Pitman Publishing, London (1976). *(pp.200,201)*

J3 E. Jenckel, *On the determination of the temperature intervals for transition in glass by the alteration of viscosity with temperature.* Zeitschr. anorg. allgem. Chem., 216 (1934), 367-375. *(p.5)*

J4 E. Jenckel, *The effect of cooling on the properties of glasses and plastics.* Zeitschr. Elektrochem., 43 (1937), 796-806. *(pp.5,6)*

J5 E. Jenckel and K. Uberreiter, *On polystyrene glasses of various chain lengths.* Zeitschr. Phys. Chem. A, 182 (1938), 361-383. *(pp.6,82)*

E. Jenckel [T5].

J6 J. Jochims, *On the spinnability of physiological fluids and on methods of its measurement.* Koll. Z., 61 (1932), 250-256. *(p.10)*

J7 M. W. Johnson Jr. and D. Segalman, *A model for viscoelastic fluid behaviour which allows non-affine deformation.* J. Non-Newtonian Fluid Mech., 2 (1977), 255-270. *(pp.28,30)*

M. W. Johnson Jr. [B17].

J8 D. F. Jones and L. R. G. Treloar, *The properties of rubber in pure homogeneous strain.* J. Phys. D., Appl. Phys., 8 (1975), 1285-1304. *(pp.86,92)*

J9 D. W. Jordan and P. Smith, *Nonlinear ordinary differential equations.* Oxford University Press (1977). *(p.214)*

J10 D. D. Joseph, *Slow motion and viscometric motion; stability and bifurcation of the rest state of a simple fluid. Part III. Die swell - the final diameter of a capillary jet.* Arch. Rat. Mech. Anal., 56 (1974), 99-157. *(p.189)*

J11 D. D. Joseph, *Stability of fluid motions I, II.* Springer-Verlag, Berlin, (1976). *(p.205)*

J12 D. D. Joye, G. W. Poehlein and C. D. Denson, *A bubble inflation technique for the measurement of viscoelastic properties in equal biaxial extensional flow.* Trans. Soc. Rheol., 16 (1972), 421-445. *(pp.88,91,159)*

J13 D. D. Joye, G. W. Poehlein and C. D. Denson, *A bubble inflation technique for the measurement of viscoelastic properties in equal biaxial extensional flow - II.* Trans. Soc. Rheol., 17 (1973), 287-301. *(pp.88,89,91,159)*

K1 E. J. Kaltenbacher, C. B. Howard and H. D. Parson, *The use of melt strength in predicting the processability of polyethylene extrusion coating resins.* Tappi, 50 (1967), 20-26. *(p.103)*

K2 E. Kamei and S. Onogi, *Extensional and fractural properties of monodisperse polystyrenes at elevated temperatures.* Appl. Polym. Symp., 27 (1975), 19-46. *(p.82)*

K3 F. A. Kanel, *The extension of viscoelastic materials.* Ph.D. dissertation, University of Delaware (1972). *(pp.100,103)*

K4 H. J. Karam and J. C. Bellinger, *Tensile creep of polystyrene at elevated temperatures. Part I.* Trans. Soc. Rheol., 8 (1964), 61-72. *(p.82)*

K5 H. J. Karam, K. S. Hyun and J. C. Bellinger, *Rheological properties of molten polymers. II. Creep function of commercial polystyrene.* Trans. Soc. Rheol., 13 (1969), 209-229. *(p.82)*

K6 V. A. Kargin and T. I. Sogolova, *On the question of the three physical states of amorphous linear polymers.* Zh. Fiz. Khim., 23 (1949), 530-539. *(p.77)*

K7 V. A. Kargin and T. I. Sogolova, *Development of a method of study of the true processes of flow in polymers.* Zh. Fiz. Khim., 23 (1949), 540-550. *(p.6)*

K8 V. A. Kargin and T. I. Sogolova, *Investigation of the process of viscous flow of polyisobutylene.* Zh. Fiz. Khim., 23 (1949), 551-562. *(pp.6,84)*

K9 S. Kase and T. J. Matsuo, *Studies on melt spinning. I. Fundamental equations on the dynamics of melt spinning.* J. Polym. Sci. A, 3 (1965), 2541-2554. *(pp.93,97,102)*

K10 S. Kase and T. Matsuo, *Studies on melt spinning. II. Steady state and transient solutions of fundamental equations compared with experimental results.* J. Appl. Polym. Sci., 11 (1967), 251-287. *(pp.97,102)*

K11 S. Kase, *Studies on melt spinning. III. Velocity field within the thread.* J. Appl. Polym. Sci., 18 (1974), 3267-3278. *(p.198)*

S. Kase [I6].

K12 A. Kaye, *Non-Newtonian flow in incompressible fluids.* College of Aeronautics, Cranfield, Tech. Note 134 (1962). *(p.34)*

K13 A. Kaye and D. G. Vale, *The shape of a vertically falling stream of a Newtonian liquid.* Rheol. Acta, 8 (1969), 1-5. *(pp.193, 197)*

K14 Y. Kazama, Y. Sumida, H. Ogawa, S. Okai and T. Hiraoka, *Mechanical behaviour of elastomers under deformation/tensile resilience.* VIIth Int. Cong. Rheol., [K15] (1976), 466-467. *(p.86)*

E. A. Kearsley [B11].

K. Kedzierska [Z7].

A. Keller [M6].

W. D. Kennedy [W4].

O.-K. Kim [L13].

Y. W. Kim [H3].

M. J. King [W6].

K15 C. Klason and J. Kubat, editors, *Proceedings of the VIIth International Congress on Rheology.* Swedish Society of Rheology, c/o Department of Polymeric Materials, Chalmers University of Technology, Gothenburg (1976). *(p.214)*

K16 P. Kohler, *Accurate determination of temperature and diameter of a melt spun fiber.* Chem. Ing. Tech., 43 (1971), 274-279. *(p.103)*

K17 O. Kramer, R. Greco, J. D. Ferry and E. T. McDonel, *Rubber networks containing unattached macromolecules. II. Viscoelastic properties in small strains of the system Ethylene-Propylene Terpolymer with Ethylene-Propylene Copolymer.* J. Polym. Sci., Polym. Phys., 13 (1975), 1675-1685. *(p.86)*

O. Kramer [G19,T14].

J. Kubat [K15].

J. S. Y. Lai [F7].

L1 M. O. Lai and D. L. Holt, *The extensional flow of polymethylmethacrylate and high-impact polystyrene at thermoforming temperatures.* J. Appl. Polym. Sci., 19 (1975), 1209-1220. *(pp.82,85)*

L2 M. O. Lai and D. L. Holt, *Thickness variation in the thermoforming of PMMA and high-impact polystyrene sheets.* J. Appl. Polym. Sci., 19 (1975), 1805-1814. *(p.92)*

R. R. Lamonte [H2].

H. Lamp [T16,T17].

R. F. Landel [P7,P8].

L3 H. M. Laun and H. Münstedt, *Comparison of the elongational behaviour of a polyethylene melt at constant stress and constant strain rate.* Rheol. Acta, 15 (1976), 517-524. *(pp.72, 73,79,81,83,156)*

L4 H. M. Laun, *Elongational behaviour of polyethylene melts at constant strain rate.* Proc. Brit. Soc. Rheol. Conf., [F5] (1977), 36-38. *(pp.73,79,80,81,83,84)*

L5 H. M. Laun and H. Münstedt, *Elongational behaviour of a LDPE melt. I. Strain rate and stress dependence of viscosity and recoverable strain in the steady state. Comparison with shear data. Influence of interfacial tension.* Rheol. Acta, 17 (1978), 415-425. *(pp.74,75,157,158)*

N. Laws [G21].

L6 H. Leaderman, *Large longitudinal retarded elastic deformation of rubber-like network polymers.* Trans. Soc. Rheol., 6 (1962). 361-382. *(pp.6,72,86)*

L7 W.-K. Lee and H. H. George, *Flow visualization of fiber suspensions.* Polym. Eng. Sci., 18 (1978), 146-156. *(p.211)*

W.-K. Lee [A11,R7].

L8 P. J. Leider, *Squeezing flow between parallel disks. II. Experimental results.* Ind. Eng. Chem. Fundam., 13 (1974), 342-346. *(p.211)*

L9 A. I. Leonov, *Nonequilibrium thermodynamics and rheology of viscoelastic polymer media.* Rheol. Acta, 15 (1976), 85-98. *(p.7)*

L10 A. I. Leonov, E. H. Lipkina, E. D. Pashkin and A. N. Prokunin, *Theoretical and experimental investigation of shearing in elastic polymer liquids.* Rheol. Acta, 15 (1976), 411-426. *(p.7)*

A. I. Leonov [G17,V1].

Y. K. Li [C11].

G. Lianis [F11,G13,M3,M4].

S. Lidorikis [V11].

L11 H. R. Lillie, *Viscosity of glass between the strain point and the melting temperature.* J. Amer. Ceram. Soc., 14 (1931), 502-511. *(pp.4,5)*

L12 L. C. T. Lin and J. Hauenstein, *Cooling and attenuation of threadline in melt spinning of PET.* J. Appl. Polym. Sci., 18 (1974), 3509-3521. *(pp.97,102)*

E. H. Lipkina [L10].

L13 R. C. Little, R. J. Hansen, D. L. Hunston, O.-K. Kim, R. L. Patterson and R. Y. Ting, *The drag reduction phenomenon. Observed characteristics, improved agents and proposed mechanisms.* Ind. Eng. Chem. Fundam., 14 (1975), 283-296. *(p.211)*

L14 G. Locati, *A model for interpreting die swell of polymer melts.* Rheol. Acta, 15 (1976), 525-532. *(p.189)*

L15 A. S. Lodge, *Elastic liquids.* Academic Press (1964). *(pp.28,44)*

L16 A. S. Lodge and J. Meissner, *Comparison of network theory predictions with stress/time data in shear and elongation for a low-density polyethylene melt.* Rheol. Acta, 12 (1973), 41-47. *(p.156)*

L17 A. S. Lodge, *Body tensor fields in continuum mechanics (with applications to polymer rheology).* Academic Press (1974). *(pp.28, 67,120,126,127,131,134,156,213)*

L18 A. S. Lodge, J. B. McLeod and J. A. Nohel, *A nonlinear singularly perturbed Volterra integrodifferential equation occurring in polymer rheology.* Proc. Roy. Soc. Edinburgh, 80A (1978), 99-137. *(pp.144,147,148,149)*

A. S. Lodge [C7,C8].

J. M. Lorntson [A11,M8,R7].

M1 I. F. Macdonald, *Rate-dependent viscoelastic models. I. Experimental results as guidelines for selecting the memory function.* Rheol. Acta, 14 (1975), 899-905. *(p.32)*

M2 I. F. Macdonald, *Rate-dependent viscoelastic models. II. The MBC-model: an experimental assessment.* Rheol. Acta, 14 (1975), 906-918. *(p.33)*

M3 C. W. McGuirt and G. Lianis, *Experimental investigation of nonlinear viscoelasticity with variable histories.* Proc. 5th Int. Cong. Rheol., 1 (1970), 337-352. *(pp.87,92)*

M4 C. W. McGuirt and G. Lianis, *Constitutive equations for viscoelastic solids under finite uniaxial and biaxial deformations.* Trans. Soc. Rheol., 14 (1970), 117-134. *(pp.87,92)*

M5 G. R. McKay, J. Ferguson and N. E. Hudson, *Elongational flow and the wet spinning process.* J. Non-Newtonian Fluid Mech., 4 (1978), 89-98. *(pp.99,104)*

P. Mackie [H13].

D. R. McLaren [J1].

J. B. McLeod [L18].

M6 M. R. Mackley and A. Keller, *Flow induced polymer chain extension and its relation to fibrous crystallisation.* Phil. Trans. Roy. Soc., A278 (1975), 29-66. *(p.111)*

M7 M. R. Mackley, *Flow singularities, polymer chain extension and hydrodynamic instabilities.* J. Non-Newtonian Fluid Mech., 4 (1978), 111-136. *(pp.56,108,111,112,114,116)*

M. R. Mackley [B12].

M8 C. W. Macosko and J. M. Lorntson, *The rheology of two blow moulding polyethylenes.* SPE Tech. Papers, 19 (1973), 461-467. *(pp.74,84)*

W. C. MacSporran [G18].

M9 J. M. Maerker and W. R. Schowalter, *Biaxial extension of an elastic liquid.* Rheol. Acta, 13 (1974), 627-638. *(pp.88,89,91)*

A. Ya. Malkin [V3,V8].

F. H. Malpress [S4].

F. P. La Mantia [A4,A5,A6,A7,A8,A9].

M10 G. Marrucci and G. Astarita, *Linear steady, two-dimensional flows of viscoelastic liquids.* AIChE J., 13 (1967), 931-935. *(pp.112,114)*

M11 G. Marrucci and R. E. Murch, *Steady symmetric sink flows of incompressible simple fluids.* Ind. Eng. Chem. Fundam., 9 (1970), 498-500. *(p.115)*

M12 G. Marrucci, G. Titomanlio and G. C. Sarti, *Testing a constitutive equation for entangled networks by elongational and shear data of polymer melts.* Rheol. Acta, 12 (1973), 269-275. *(p.50)*

M13 G. Marrucci, *Limiting concepts in extensional flow.* Polym. Eng. Sci., 15 (1975), 229-233. *(pp.111,112)*

M14 G. Marrucci and A. Ciferri, *Phase equilibria of rod-like molecules in an extensional flow field.* J. Polym. Sci.: Polymer Letters, 15 (1977), 643-648. *(p.211)*

G. Marrucci [A4,A5,A7,A9,A17,D10,D13].

D. E. Marshall [C20].

M15 W. R. Marshall Jr. and R. L. Pigford, *The application of differential equations to chemical engineering problems.* University of Delaware Press (1947). *(pp.9,192)*

S. G. Maskell [C20].

M16 M. A. Matovich, *Mechanics of a spinning threadline.* Ph.D. dissertation, University of Cambridge (1966). *(p.177)*

M17 M. A. Matovich and J. R. A. Pearson, *Spinning a molten threadline - steady state isothermal viscous flows.* Ind. Eng. Chem. Fundam., 8 (1969), 512-520. *(pp.163,171,184,192,196,197)*

M. A. Matovich [P3].

M18 M. Matsui and D. C. Bogue, *Non-isothermal rheological response in melt spinning and idealised elongational flow.* Polym. Eng. Sci., 16 (1976), 735-741. *(pp.97,102,183,187)*

T. J. Matsuo [K9,K10].

M19 J. Meissner, *A rheometer for investigation of deformation-mechanical properties of plastic melts under defined extensional straining.* Rheol. Acta, 8 (1969), 78-88. *(pp.73,83)*

M20 J. Meissner, *Elongational properties of polyethylene melts.* Rheol. Acta, 10 (1971), 230-242. *(pp.73,77,79,83,103)*

M21 J. Meissner, *Development of a universal extensional rheometer for the uniaxial extension of polymer melts.* Trans. Soc. Rheol., 16 (1972), 405-420. *(pp.72,73,77,83,103)*

M22 J. Meissner, *Modifications of the Weissenberg rheogoniometer for measurement of transient rheological properties of molten PE under shear. Comparison with tensile data.* J. Appl. Polym. Sci., 16 (1972), 2877-2899. *(p.79)*

M23 J. Meissner, *Basic parameters, melt rheology, processing and end-use properties of three similar low density polyethylene samples.* Pure Appl. Chem., 42 (1975), 553-612. *(pp.83,103)*

M24 J. Meissner, *Combined constant shear rate and stress relaxation test for linear viscoelastic fluids.* J. Polym. Sci.: Polym. Phys., 16 (1978), 915-919. *(p.142)*

J. Meissner [L16].

M25 B. J. Meister, *An integral constitutive equation based on molecular network theory.* Trans. Soc. Rheol., 15 (1971), 63-89. *(pp.31,48)*

M26 A. B. Metzner, E. A. Uebler and C. F. Chan Man Fong, *Converging flows of viscoelastic materials.* AIChE J., 15 (1969), 750-758. *(p.207)*

M27 A. B. Metzner and A. P. Metzner, *Stress levels in rapid extensional flows of polymeric fluids.* Rheol. Acta, 9 (1970), 174-181. *(p.56)*

M28 A. B. Metzner, *Extensional primary field approximations for viscoelastic media.* Rheol. Acta, 10 (1971), 434-444. *(pp.209,210,211)*

A. B. Metzner [A11,B1,M29,R7,S7,S15,T20,W10].

A. P. Metzner [M27].

M29 J. Mewis and A. B. Metzner, *The rheological properties of suspensions of fibres in Newtonian fluids subjected to extensional deformation.* J. Fluid Mech., 62 (1974), 593-600. *(pp.18,95,104,106,211)*

M30 S. Middleman, *Stability of a viscoelastic jet.* Chem. Eng. Sci., 20 (1965), 1037-40. *(p.18)*

M31 S. Middleman, *Transient response of an elastomer to large shearing and stretching deformations.* Trans. Soc. Rheol., 13 (1969), 123-139. *(p.84)*

M32 S. Middleman, *Fundamentals of polymer processing.* McGraw Hill, New York (1977). *(p.213)*

S. Middleman [P1].

M33 C. A. Moore and J. R. A. Pearson, *Experimental investigations into an isothermal spinning threadline: Extensional rheology of a Separan AP30 solution in glycerol and water.* Rheol. Acta, 14 (1975), 436-446. *(pp.96,101,104)*

D. R. Moore [C26].

D. H. Morris [S1].

M34 F. H. Muller, *Fibre formation, fibre properties and molecular structure.* Physik. Zeitschr., 42 (1941), 123-129. *(p.12)*

M35 F. H. Muller and G. Binder, *Nonlinear deformation. II. Cold drawing.* Koll.-Z., 183 (1962), 120-134. *(p.7)*

M36 F. H. Muller and C. Engelter, *On the stress-dependence of flow of polymers.* Koll.-Z., 186 (1962), 36-41. *(p.7)*

F. H. Muller [B14].

M37 H. Münstedt, *Viscoelasticity of polystyrene melts in tensile creep experiments.* Rheol. Acta, 14 (1975), 1077-1088. *(pp.72,82)*

H. Münstedt [L3,L5].

R. E. Murch [M11].

N1 P. M. Naghdi, *A direct formulation of Newtonian and non-Newtonian fluid jets.* IUTAM abstracts, [C38] (1978). *(p.199)*

F. J. Van Natta [C2].

N2 F. Nazem and C. T. Hill, *Elongational and shear viscosities of a bead-filled thermoplastic.* Trans. Soc. Rheol., 18 (1974), 87-101. *(pp.72,87)*

N3 H. Nguyen and D. V. Boger, *The kinematics and stability of die entry flows.* IUTAM preprints, [C38] (1978). *(p.207)*

N4 R. E. Nickell, R. I. Tanner and B. Caswell, *The solution of viscous incompressible jet and free-surface flows using finite-element methods.* J. Fluid Mech., 65 (1974), 189-206. *(p.189)*

N5 L. Nicodemo, B. de Cindio and L. Nicolais, *Elongational viscosity of microbead suspensions.* Polym. Eng. Sci., 15 (1975), 679-683. *(p.103)*

L. Nicodemo [A3,A8,A16,C17].

L. Nicolais [C17,N5].

N6 C. J. van Nieuwenburg, *Viscosity and plasticity from a technical point of view,* in *First Report on Viscosity and Plasticity.* Academy of Sciences, Amsterdam, ch. IV (1935), 141-172. *(p.5)*

A. H. Nissan [G2].

N7 H. Nitschmann and J. Schrade, *On the fibre-forming ability of non-Newtonian liquids.* Helv. Chim. Acta, 31 (1948), 297-319. *(pp.12,13)*

N8 H. Nitschmann, *The viscosity anomaly causing the spinning of liquids. Stress-flow curves of liquid threads.* Proc. 1st. Int. Cong. Rheol. (1949), II-32 - II-34. *(pp.12,13)*

N9 H. Nitschmann and W. Aeschlimann, *On the problem of thread pulling in fluids - measurement of flow resistance in stretched fluid threads.* Angew. Chem., 65 (1953), 261-262. (See W. Aeschlimann, dissertation, Bern (1952).) *(p.13)*

J. A. Nohel [L18].

N10 W. Noll, *Motions with constant stretch history.* Arch. Rat. Mech. Anal., 11 (1962), 97-105. *(pp.134,135)*

W. Noll [C30,C31].

N11 J. W. M. Noordermeer and J. D. Ferry, *Nonlinear relaxation of stress and birefringence in simple extension of 1,2-polybutadiene.* J. Polym. Sci.: Polym. Phys., 14 (1976), 509-520. *(p.87)*

B. J. O'Brien [E1].

H. Ogawa [K14].

S. Okai [K14].

O1 J. G. Oldroyd, *Non-Newtonian effects in steady motion of some idealised elasticoviscous liquids.* Proc. Roy. Soc., A245 (1958), 278-297. *(pp.29,39)*

O2 J. G. Oldroyd, *The hydrodynamics of materials whose rheological properties are complicated.* Rheol. Acta, 1 (1961), 337-344. *(p.29)*

O3 J. G. Oldroyd, *Some steady flows of the general elasticoviscous liquid.* Proc. Roy. Soc., A283 (1965), 115-133. *(p.134)*

O4 D. R. Oliver and R. Bragg, *The triple jet: a new method for measurement of extensional viscosity.* Rheol. Acta, 13 (1974), 830-835. *(pp.96,104)*

O5 D. R. Oliver, *Development of the 'triple jet' system for extensional flow measurements on polymer solutions.* Colloques Int. du C.N.R.S., 233 (1975), 325-330. *(pp.99,100,104)*

O6 D. R. Oliver, *The triple jet: high strain-dependent stresses are developed in stretching polymer solutions.* Chem. Eng. J., 9 (1975), 255-256. *(pp.99,104)*

O7 D. R. Oliver and R. C. Ashton, *The triple jet: influence of shear history on the stretching of polymer solutions.* J. Non-Newtonian Fluid Mech., 1 (1976), 93-104. *(pp.104,188)*

D. R. Oliver [W6].

S. Onogi [K2].

K. Orii [S2].

Y. Oyanagi [C5].

J. Y. Park [H4].

H. D. Parson [K1].

E. D. Pashkin [L10].

R. L. Patterson [L13].

P1 G. Pearson and S. Middleman, *Elongational flow behaviour of viscoelastic liquids: Part I. Modeling of bubble collapse. Part II. Definition and measurement of apparent elongational viscosity.* AIChE J., 23 (1977), 714-723. *(p.124)*

P2 G. H. Pearson, *personal communication* (1978). *(pp.74,83)*

G. H. Pearson [G1].

P3 J. R. A. Pearson and M. A. Matovich, *On spinning a molten threadline-stability*. Ind. Eng. Chem. Fundam., 8 (1969), 605-609. *(pp.157,205)*

P4 J. R. A. Pearson and C. J. S. Petrie, *The flow of a tubular film. Part 1. Formal mathematical representation*. J. Fluid Mech., 40 (1970), 1-19. *(pp.90,198,210)*

P5 J. R. A. Pearson, *Instability in Non-Newtonian flow*. Ann. Revs. Fluid Mech., 8 (1976), 163-181. *(pp.11,16,205,210)*

P6 J. R. A. Pearson and P. A. Gutteridge, *Stretching flows for thin film production. 1. Bubble blowing in the solid phase*. J. Non-Newtonian Fluid Mech., 4 (1978), 57-72. *(pp.159, 210)*

J. R. A. Pearson [H19,M17,M33].

P7 S. T. J. Peng and R. F. Landel, *Response of bulk polymer under motion with constant stretch histories*. Rheol. Acta, 13 (1973), 548-556. (See also Peng, JPL Quart. Rev., 2 (1972), 40-45.) *(pp.72,84,91)*

P8 S. T. J. Peng and R. F. Landel, *Concentration and temperature effects on elongational viscosity of dilute polymer solutions*. Abstract E1, 47th Ann. Meeting, Soc. of Rheology, New York (1977). (See also Peng and Landel, J. Appl. Phys., 47 (1976), 4255-4260.) *(p.103, frontispiece)*

P9 M. G. N. Perera and K. Walters, *Long-range memory effects in flows involving abrupt changes in geometry. Part 2. The expansion/contraction/expansion problem*. J. Non-Newtonian Fluid Mech., 2 (1977), 191-204. *(p.207)*

P10 A. Peterlin, *Hydrodynamics of linear macromolecules*. Pure Appl. Chem., 12 (1966), 563-586. *(p.105)*

P11 A. Peterlin, *Hydrodynamics of macromolecules in a velocity field with longitudinal gradient*. J. Polym. Sci., B, 4 (1966), 287-291. *(p.211)*

P12 C. J. S. Petrie, *A comparison of theoretical prediction with published experimental measurements in the blown film process*. AIChE J., 21 (1975), 275-282. *(pp.159,210)*

P13 C. J. S. Petrie, *The analysis of polymer processing operations involving free surface elongational flows*. Polymer Rheology and Plastics Processing, [C35] (1975), 307-319. *(p.210)*

P14 C. J. S. Petrie and M. M. Denn, *Instabilities in polymer processing*. AIChE J., 22 (1976), 209-236. *(pp.16,18,19,205,210)*

P15 C. J. S. Petrie, *Some problems in unsteady flow for co-rotational rheological models*. VIIth Int. Cong. Rheol., [K15] (1976), 446-447. *(pp.26,150,154,164,165)*

P16 C. J. S. Petrie, *On stretching Maxwell models*. J. Non-Newtonian Fluid

P16 C. J. S. Petrie - *continued*
Mech., 2 (1977), 221-253. *(pp.44,79,82,136,152,155,159,179)*

P17 C. J. S. Petrie, *Relationships between some recent integral and rate-type constitutive equations.* Proc. Brit. Soc. Rheol. Conf., [F5] (1977), pp.72-76. (Full text in a Departmental Report.) *(pp.26,183)*

P18 C. J. S. Petrie, *Some results in the theory of melt spinning for model viscoelastic liquids.* J. Non-Newtonian Fluid Mech., 4 (1978), 137-159. *(pp.26,99,140,156,169,184,187,190,191)*

P19 C. J. S. Petrie, editor, *Stretching flows.* J. Non-Newtonian Fluid Mech., 4 (1978), Issues 1-2, Elsevier, Amsterdam. *(pp.159,214)*

P20 C. J. S. Petrie, *Measures of deformation and convected derivatives.* IUTAM preprints, [C38] (1978). *(pp.25,26,28,30,47)*

C. J. S. Petrie [D12,F5,P4,W7].

R. Pfeffer [G14].

P21 Nhan Phan-Thien and R. I. Tanner, *A new constitutive equation derived from network theory.* J. Non-Newtonian Fluid Mech., 2 (1977), 353-365. *(pp.28,48,49,50)*

P22 Nhan Phan-Thien, *A nonlinear network viscoelastic model.* J. Rheol., 22 (1978), 259-283. *(pp.28,48,49,50,183)*

W. Philippoff [G3].

P23 M. C. Phillips, *The start-up of steady elongational flow of viscoelastic materials.* Theoretical Rheology, [H19] (1975), 276-291. *(p.126)*

P24 M. C. Phillips, *The prediction of time-dependent nonlinear stresses in viscoelastic materials. I. Selection of constitutive equation and strain measure.* J. Non-Newtonian Fluid Mech., 2 (1977), 109-121. *(p.47)*

J. M. Pierrard [C21,C22].

R. L. Pigford [M15].

G. Pilate [C37].

P25 A. C. Pipkin, *Approximate constitutive equations,* in Modern Developments in the Mechanics of Continua, ed. S. Eskinazi. Academic Press (1966), 89-108. *(p.35)*

P26 A. C. Pipkin, *Lectures on viscoelasticity theory.* Springer (1972). *(p.213)*

P27 A. C. Pipkin and R. I. Tanner, *A survey of theory and experiment in viscometric flows of viscoelastic liquids,* in Mechanics Today, ed. S. Nemat-Nasser, 1 (1972), 262-321. *(pp.14,213)*

P28 A. C. Pipkin and R. I. Tanner, *Steady non-viscometric flows of viscoelastic liquids.* Ann. Revs. Fluid Mech., 9 (1977), 13-32. *(p.213)*

J. F. T. Pittman [H6].

G. W. Poehlein [J12,J13].

P29 T. Poston and I. N. Stewart, *Catastrophe theory and its applications.* Pitman Publishing, London (1978). *(pp.56,214)*

P30 D. C. Pountney, *The extensional flow of non-Newtonian fluids.* Ph.D. thesis, University of Wales (1976). *(p.45)*

P31 D. C. Pountney and K. Walters, *Nearly extensional flows.* Phys. Fluids, 21 (1978), 1482-1484. *(p.209)*

A. N. Prokunin [L10,V1].

R1 B. V. Radushkevich, V. D. Fikhman and G. V. Vinogradov, *Uniaxial uniform-speed elongation of high-elasticity liquids (of low-molecular polyisobutylene as an example).* Dokl. Phys. Chem., 180 (1968), 358-361. *(pp.72,78,84)*

B. V. Radushkevich [V2,V3,V4].

S. Ranaudo [C17].

R2 M. Reiner and A. Freudenthal, *Failure of a material showing creep. (A dynamical theory of strength).* Proc. 5th Int. Cong. Appl. Mech. (1938), 228-233. *(p.17)*

R3 M. Reinger, *The coefficient of viscous traction.* Amer. J. Math. 68 (1946), 672-680. *(p.9)*

R4 M. Reiner, *Deformation, strain and flow.* H. K. Lewis, London (1960). *(p.17)*

R5 O. Reynolds, *On the theory of lubrication and its application to Mr. Beauchamp Tower's experiments, including an experimental determination of the viscosity of olive oil.* Phil. Trans. Roy. Soc. A177 (1886), 157-234. *(p.7)*

R6 J. Rhi-Sausi and J. M. Dealy, *An extensiometer for molten plastics.* Polym. Eng. Sci., 16 (1976), 799-802. *(pp.72,73,83)*

J. Rhi-Sausi [D6,D8].

R7 C. I. Richardson, J. M. Lorntson, W. K. Lee and A. B. Metzner, *Rheological behaviour of molten polymers in shearing and in extensional flows.* VIIth Int. Cong. Rheol., [K15] (1976), 190-191. *(p.73)*

C. I. Richardson [A11].

R8 R. S. Rivlin, *Large elastic deformations of isotropic materials. II. Some uniqueness theorems for pure, homogeneous deformation.* Phil. Trans. Roy. Soc., A240 (1948), 491-508. *(p.151)*

R9 R. S. Rivlin and S. W. Saunders, *Large elastic deformations of isotropic materials. VII. Experiments on the deformation of rubber.* Proc. Roy. Soc., A243 (1951), 251-288. *(pp.86,88,92)*

R. S. Rivlin [G20].

G. Rizzo [A5].

J. M. Rodriguez [A1].

J. F. Roman [W14].

R10 M. Ronay, *Determination of the dynamic surface tension of liquids from the growth rate of axisymmetric disturbances on an excited capillary jet.* VIIth Int. Cong. Rheol., [K15] (1976), 210-211. *(p.19)*

R11 G. Ronca, *A network theory of isothermal spinning.* Rheol. Acta, 15 (1976), 628-641. *(pp.183,205)*

R12 R. Roscoe, *The steady elongation of elasto-viscous liquids.* Brit. J. Appl. Phys., 16 (1965), 1567-1571. *(p.120)*

P. Le Roy [C22].

W. Ruland [B9].

S1 M. H. Sadd and D. H. Morris, *Rate-dependent stress-strain behaviour of polymeric materials.* J. Appl. Polym. Sci., 20 (1976), 421-433. *(p.85)*

S2 Y. Sano, K. Orii and N. Yamada, *Trouton viscosity of polypropylene in the melt spinning process.* Sen-I Gakkaishi, 24 (1968), 147-154. *(pp.97,98,102)*

G. C. Sarti [M12].

D. W. Saunders [R9].

S3 L. R. Schmidt and J. F. Carley, *Biaxial stretching of heat-softened plastic sheets: experiments and results.* Polym. Eng. Sci., 15 (1975), 51-62. *(p.92)*

W. R. Schowalter [M9].

J. Schrade [N7].

H. Schreiner [B9].

S4 G. W. Scott-Blair, S. J. Folley, F. H. Malpress and F. M. V. Coppen, *Variation in certain properties of bovine cervical mucus during the oestrus cycle.* Biochem. J., 35 (1941), 1039-1049. *(p.11)*

S5 G. W. Scott-Blair, *A survey of general and applied rheology.* Pitman, London (1944). *(p.5)*

O. Scrivener [C21].

D. Segalman [J7].

S6 M. T. Shaw, *Extensional viscosity of melts using a programmable tensile testing machine.* VIIth Int. Cong. Rheol., [K15] (1976), 304-305. *(pp.73,83,84)*

S7 R. E. Sheffield and A. B. Metzner, *Flows of nonlinear fluids through porous media.* AIChE J., 22 (1976), 736-744. *(p.212)*

R. Shinnar [G14,G15].

S8 R. Signer, *Problems of artificial fibre preparation.* Chemiker-Zeitung,

S8 R. Signer - *continued*
79 (1955), 332-335, 371-373. *(pp.12,13)*

J. M. Simmons [T7].

S9 J. C. Slattery, *Unsteady relative extension of incompressible simple fluids*. Phys. Fluids, 7 (1964), 1913-1914. *(p.126)*

S. Smale [H11].

P. Smith [J9].

S10 T. L. Smith, *Viscoelastic behaviour of polyisobutylene under constant rates of elongation*. J. Polym. Sci., 20 (1956), 89-100. *(p.84)*

S11 T. L. Smith, *Dependence of the ultimate properties of a GR-S rubber on strain rate and temperature*. J. Polym. Sci., 32 (1958), 99-113. *(p.87)*

S12 T. L. Smith and J. E. Frederick, *Viscoelastic properties of a styrene-butadiene vulcanizate in large biaxial and simple tensile deformations*. Trans. Soc. Rheol., 12 (1968), 363-396. *(pp.87,88,92)*

S13 T. L. Smith and R. A. Dickie, *Effect of finite extensibility on the viscoelastic properties of a styrene-butadiene rubber vulcanizate in simple tensile deformations up to rupture*. J. Polym. Sci., A2, 7 (1969), 635-658. *(p.87)*

S14 T. L. Smith, *Linear viscoelastic response to a deformation at constant rate: derivation of physical properties of a densely cross-linked elastomer*. Trans. Soc. Rheol., 20 (1976), 103-117. *(p.86)*

T. L. Smith [D21].

T. I. Sogolova [K6,K7,K8].

D. H. Solomon [E1].

S15 J. A. Spearot and A. B. Metzner, *Isothermal spinning of molten polyethylenes*. Trans. Soc. Rheol., 16 (1972), 495-518. *(pp.48,98,100,102,183)*

S16 T. W. Spriggs, J. D. Huppler and R. B. Bird, *An experimental appraisal of viscoelastic models*. Trans. Soc. Rheol., 10 (1966), 191-213. *(p.28)*

J. E. Spruiell [B7,D9].

S17 J. Stefan, *Experiments on apparent adhesion*. Sitzber. K. Akad. Wiss. Wien II, 69 (1874), 713-735. *(p.7)*

W. Stehrenberger [T9].

S18 J. F. Stevenson, *Elongational flow of polymer melts*. AIChE J., 18 (1972), 540-547. *(pp.72,84,86,122)*

S19 J. F. Stevenson, S. C.-K. Chung and J. T. Jenkins, *Evaluation of material functions for steady elongational flows*. Trans. Soc. Rheol., 19 (1975), 397-405. *(p.120)*

J. F. Stevenson [B17,C16].

I. N. Stewart [P29].

S20 V. H. Stott, *The viscosity of glass.* J. Soc. Glass Technol. Trans., 9 (1925), 207-225. *(p.4)*

S21 K. Strauss, *The flow of a simple viscoelastic liquid in a converging channel. I. Steady flow.* Acta Mechanica, 20 (1974), 233-246. *(p.207)*

S22 K. Strauss, *The flow of a simple viscoelastic liquid in a converging channel. II. Stability of the flow.* Acta Mechanica, 21 (1975), 141-152. *(p.207)*

Y. Sumida [K14].

T1 T. Takaki and D. C. Bogue, *The extensional and failure properties of polymer melts.* J. Appl. Polym. Sci., 19 (1975), 419-433. *(p.83)*

T2 R. Takserman-Krozer, *Behaviour of polymer solutions in a velocity field with parallel gradient; III, IV.* J. Polym. Sci. A, 1 (1963), 2477-2494. *(pp.105,211)*

T3 R. Takserman-Krozer, *Hydrodynamic properties of dilute polymer solutions in a general velocity field.* J. Polym. Sci. C, 16 (1967), 2855-2865. *(pp.208,211)*

T4 G. Tammann and R. Tampke, *On the spinnability, surface tension and specific heat of glasses.* Z. anorg. allgem. Chem., 162 (1927), 1-16. *(p.10)*

T5 G. Tammann and E. Jenckel, *On the rate of stretching of heated glass fibres.* Z. anorg. allgem. Chem., 191 (1930), 122-127. *(pp.4,5)*

R. Tampke [T4].

T6 R. I. Tanner, *Some illustrative problems in the flow of viscoelastic non-Newtonian lubricants.* A.S.L.E. Trans., 8 (1965), 179-183. *(p.47)*

T7 R. I. Tanner and J. M. Simmons, *Combined simple and sinusoidal shearing in elastic liquids.* Chem. Eng. Sci., 22 (1967), 1803-1815. *(p.33)*

T8 R. I. Tanner, *Comparative studies of some simple viscoelastic theories.* Trans. Soc. Rheol., 12 (1968), 155-182. *(pp.33,46)*

T9 R. I. Tanner and W. Stehrenberger, *Stresses in dilute solutions of bead-nonlinear-spring macromolecules. I. Steady potential and plane flows.* J. Chem. Phys., 55 (1971), 1958-1964. *(pp.111,112)*

T10 R. I. Tanner, *Stresses in dilute solutions of bead-nonlinear-spring macromolecules. II. Unsteady flow and approximate constitutive equations.* Trans. Soc. Rheol., 19 (1975), 37-65. *(p.111)*

T11 R. I. Tanner, *Stresses in dilute solutions of bead-nonlinear-spring macromolecules. III. Friction coefficient varying with dumbbell extension.* Trans. Soc. Rheol., 19 (1975), 557-582. *(p.106)*

T12 R. I. Tanner and R. R. Huilgol, *On a classification scheme for flow fields*. Rheol. Acta, 14 (1975), 959-962. *(pp.107,108)*

T13 R. I. Tanner, *A test particle approach to flow classification for viscoelastic fluids*. AIChE J., 22 (1976), 910-918. *(pp.110,111,112)*

R. I. Tanner [N4,P21,P27,P28,W21].

T14 C. R. Taylor, R. Greco, O. Kramer and J. D. Ferry, *Nonlinear stress relaxation of polyisobutylene in simple extension*. Trans. Soc. Rheol., 20 (1976), 141-152. *(p.85)*

C. R. Taylor [G19].

T15 G. I. Taylor, *The formation of emulsions in definable fields of flow*. Proc. Roy. Soc., A146 (1934), 501-523. *(pp.54,56)*

T16 H. Thiele and H. Lamp, *On the spinnability of colloidal solutions*. Koll.-Z., 129 (1952), 25-39. *(p.10)*

T17 H. Thiele and H. Lamp, *On the spinnability of colloidal systems*. Koll.-Z., 173 (1960), 63-72. *(p.10)*

T18 D. W. Thompson, *On growth and form* (abridged edition). Cambridge University Press (1961). *(pp.10,11)*

T19 R. Y. Ting and D. L. Hunston, *Some limitations on the detection of high elongational stress effects in dilute polymer solutions*. J. Appl. Polym. Sci., 21 (1977), 1825-1833. *(p.155)*

R. Y. Ting [L13].

T20 G. Titomanlio, B. E. Anshus, G. Astarita and A. B. Metzner, *The rheology of solid polymers subjected to large deformations*. Trans. Soc. Rheol., 20 (1976), 527-543. *(p.85)*

G. Titomanlio [A2,A3,A4,A5,A6,M12].

R. W. Tock [H12].

T21 B. A. Toms, *Elastic and viscous properties of dilute solutions of PMMA in certain solvent/non-solvent mixtures*. Rheol. Acta, 1 (1958), 137-141. *(p.10)*

T22 L. R. G. Treloar, *Strains in an inflated rubber sheet, and the mechanism of bursting*. Trans. I.R.I., 19 (1944), 201-212. *(pp.88, 92)*

L. R. G. Treloar [J8].

T23 F. T. Trouton and E. S. Andrews, *On the viscosity of pitch-like substances*. Phil. Mag. (Ser. 6), 7 (1904), 347-355. *(pp.2,3)*

T24 F. T. Trouton, *On the coefficient of viscous traction and its relation to that of viscosity*. Proc. Roy. Soc., A77 (1906), 426-440. *(pp.2,3,8,11,18,192)*

T25 C. Truesdell, *A first course in rational continuum mechanics. Volume 1 - General concepts*. Academic Press, New York (1977). *(pp.128,129,132,134,213)*

N. W. Tschoegl [C10].

W. E. S. Turner [B24].

K. Uberreiter [J5].

E. A. Uebler [M26].

L. Utracki [D6].

D. G. Vale [K13].

V1 G. V. Vinogradov, A. I. Leonov and A. N. Prokunin, *On uniaxial extension of an elasto-viscous cylinder.* Rheol. Acta, 8 (1969), 482-490. *(pp.74,78,84)*

V2 G. V. Vinogradov, B. V. Radushkevich and V. D. Fikhman, *Extension of elastic liquids: polyisobutylene.* J. Polym. Sci., A2, 8 (1970), 1-17. *(pp.12,72,74,78,84,91)*

V3 G. V. Vinogradov, V. D. Fikhman, B. V. Radushkevich and A. Ya Malkin, *Viscoelastic and relaxation properties of a polystyrene melt in axial extension.* J. Polym. Sci., A2, 8 (1970), 657-678. *(pp.72,78,82)*

V4 G. V. Vinogradov, V. D. Fikhman and B. V. Radushkevich, *Uniaxial extension of polystyrene at true constant stress.* Rheol. Acta, 11 (1972), 286-291. *(pp.72,78,82)*

V5 G. V. Vinogradov, V. D. Fikhman and V. M. Alekseyeva, *Uniaxial extension of polyvinyl chloride in the high-elastic state.* Polym. Eng. Sci., 12 (1972), 317-322. *(pp.78,86)*

V6 G. V. Vinogradov, *Critical regimes of deformation of liquid polymeric systems.* Rheol. Acta, 12 (1973), 357-373. [See also Pure Appl. Chem. (suppl.) MMC-8, 413-447.] *(p.189)*

V7 G. V. Vinogradov, *Viscoelasticity and fracture phenomenon in uniaxial extension of high-molecular linear polymers.* Rheol. Acta, 14 (1975), 942-954. *(pp.78,86,87)*

V8 G. V. Vinogradov, A. Ya Malkin and V. V. Volosevitch, *Some fundamental problems in viscoelastic behaviour of polymers in shear and extension.* Appl. Polym. Symp., 27 (1975), 47-59. *(p.17)*

V9 G. V. Vinogradov, *Ultimate regimes of deformation of linear flexible-chain fluid polymers.* Polymer, 18 (1977), 1275-1285. *(p.17)*

V10 G. V. Vinogradov, *Overshoot in elongation at constant strain rate - comparison of data of Meissner and Vinogradov.* Personal communication (1977). *(p.78)*

G. V. Vinogradov [R1].

V11 J. Vlachopoulos, M. Horie and S. Lidorikis, *An evaluation of expressions predicting die swell.* Trans. Soc. Rheol., 16 (1972), 669-685. *(p.189)*

V. V. Volosevitch [V8].

V12 A. J. De Vries and C. Bonnebat, *Uni- and biaxial stretching of chlorinated PVC sheets. A fundamental study of thermoformability.* Polym. Eng. Sci., 16 (1976), 93-100. *(pp.73,86,88,89,92)*

V13 A. J. De Vries, C. Bonnebat and J. Beautemps, *Uni- and biaxial orientation of polymer fibres and sheets.* J. Polym. Sci.: Polym. Symp., 58 (1977), 109-156. *(pp.83,85,88,89,92)*

W1 M. H. Wagner, *The film-blowing process as a rheological-thermodynamic process.* Rheol. Acta, 15 (1976), 40-51. *(pp.159,210)*

W2 M. H. Wagner, *A constitutive analysis of uniaxial elongational flow data of a low-density polyethylene melt.* J. Non-Newtonian Fluid Mech., 4 (1978), 39-55. *(pp.47,80,156)*

W3 K. Walters, *Rheometry.* Chapman and Hall, London (1975). *(pp.14,213)*

K. Walters [B25,D2,H19,P9,P31].

W4 F. C. Wampler, D. R. Gregory and W. D. Kennedy, *Spinning of poly(ethylene terephthalate) fibers from a melt pool without a spinneret.* J. Appl. Polym. Sci., 16 (1972), 99-106. *(p.96)*

W5 C. C. Wang, *A representation theorem for the constitutive equation of a simple material in motions with constant stretch history.* Arch. Rat. Mech. Anal., 20 (1965), 329-340. *(p.135)*

W6 N. D. Waters, M. J. King and D. R. Oliver, *Simple rheological models for the prediction of stretching jet profiles.* J. Non-Newtonian Fluid Mech., 2 (1977), 385-391. *(p.100)*

W7 N. D. Waters and C. J. S. Petrie, *The existence of solutions for a stretching jet.* J. Non-Newtonian Fluid Mech., 4 (1978), 161-166. *(pp.184,185)*

W8 C. B. Weinberger and J. D. Goddard, *Extensional flow behaviour of polymer solutions and particle suspensions in a spinning motion.* Int. J. Multiphase Flow, 1 (1974), 465-486. *(pp.18,95, 100,104,106,211)*

C. B. Weinberger [D24].

W9 R. A. Wessling and T. Alfrey Jr., *Stretching of elastic tubes over rotationally symmetric mandrels.* Trans. Soc. Rheol., 8 (1964), 85-100. *(p.92)*

W10 J. L. White and A. B. Metzner, *Development of constitutive equations for polymeric melts and solutions.* J. Appl. Polym. Sci., 7 (1963), 1867-1889. *(pp.47,48)*

W11 J. L. White, *A continuum theory of nonlinear viscoelastic deformation with application to polymer processing.* J. Appl. Polym. Sci., 8 (1964), 1129-1146. *(p.134)*

W12 J. L. White, *Theoretical considerations of biaxial stretching of viscoelastic fluid sheets with application to plastic sheet forming.* Rheol. Acta, 14 (1975), 600-611. *(pp.158,159)*

W13 J. L. White and Y. Ide, *Rheology and dynamics of fiber formation from polymer melts.* Appl. Polym. Symp., 27 (1975), 61-102. *(pp.95,158)*

W14 J. L. White and J. F. Roman, *Extrudate swell during the melt spinning of fibers - influence of rheological properties and take-up force.* J. Appl. Polym. Sci., 20 (1976), 1005-1023. *(p.188)*

W15 J. L. White, *Dynamics and structure development during melt spinning of fibers.* J. Soc. Rheol. Japan, 4 (1976), 137-148. *(p.12)*

W16 J. L. White, *Processability of rubber and rheological behaviour.* Rubber Chem. Technol., 50 (1977), 163-185. *(p.211)*

W17 J. L. White, *Experimental study of elongational flow of polymers.* Appl. Polym. Symp., (in press, November 1978) (also University of Tennessee, Polymer Science and Engineering Report No. 104, July 1977). *(p.71)*

W18 J. L. White and Y. Ide, *Instabilities and failure in elongational flow and melt spinning of fibres.* J. Appl. Polym. Sci., 22 (1978), 3057-3074. *(pp.16,17,18,205)*

J. L. White [A1,B7,C5,C14,I2,I3,I4].

W19 F. E. Wiley, *Working-range flow properties of thermoplastics.* Ind. Eng. Chem., 33 (1941), 1377-1380. *(pp.6,8)*

W20 G. Wilhelm, *The cooling of a melt spun polymer fibre in the spinning way.* Koll.-Z., 208 (1966), 97-]23. *(p.103)*

W21 G. Williams and R. I. Tanner, *Effect of combined shearing and stretching in viscoelastic lubrication.* Trans. A.S.M.E., Ser. F, 92 (1970), 216-219. *(p.211)*

W22 J. G. Williams, *A method for calculation for thermoforming plastics sheets.* J. Strain Anal., 5 (1970), 49-57. *(p.85)*

W23 A. S. Wineman, *Large axisymmetric inflation of a nonlinear viscoelastic membrane by lateral pressure.* Trans. Soc. Rheol., 20 (1976), 203-225. (See also J. Non-Newtonian Fluid Mech., 4 (1978), 249-260.) *(p.160)*

W24 K. Wissbrun, *Interpretation of the melt strength test.* Polym. Eng. Sci., 13 (1973), 342-345. *(pp.6,103)*

K. F. Wissbrun [A11].

G. P. Wood [G2].

N. Yamada [S2].

Y1 M. Yamamoto, *The viscoelastic properties of network structure. I. General formalism.* J. Phys. Soc. Japan, 11 (1956), 413-421. *(p.47)*

Y2 M. Yamamoto, *The viscoelastic properties of network structure. II. Structural viscosity.* J. Phys. Soc. Japan, 12 (1956), 1148-1158. *(p.47)*

Y3 M. Yamamoto, *Constitutive equations for nonlinear viscoelasticity.* Appl. Polym. Symp., 20 (1973), 3-22. *(p.47)*

J. Yerushalmi [G14,G15].

M. Yoshinari [I7].

Z1 S. Zahorski, *Flows with constant stretch history and extensional viscosity.* Arch. Mech., 23 (1971), 433-445. *(pp.121, 135)*

Z2 S. Zahorski, *Flows with proportional stretch history.* Arch. Mech., 24 (1972), 681-699. *(p.126)*

Z3 L. J. Zapas and T. Craft, *Correlation of large longitudinal deformations with different strain histories.* J. Res. Nat. Bur. Stand., 69A (1965), 541-546. *(pp.84,86)*

Z4 L. J. Zapas, *Viscoelastic behaviour under large deformations.* J. Res. Nat. Bur. Stand., 70A (1966), 525-532. *(pp.34,92)*

L. J. Zapas [B11].

Z5 G. R. Zeichner, *Spinnability of viscoelastic fluids.* M.Ch.E. dissertation, University of Delaware (1973). *(pp.102,183)*

Z6 A. Ziabicki, *Studies on the orientation phenomenon by fiber formation from polymer melts. II. Theoretical considerations.* J. Appl. Polym. Sci., 2 (1959), 24-31. *(pp.105,211)*

Z7 A. Ziabicki and K. Kedzierska, *Mechanical aspects of fibre spinning process in molten polymers. I. Stream diameter and velocity distribution along the spinning way.* Koll.-Z., 171 (1960), 51-61. *(p.103)*

Z8 A. Ziabicki, *Mechanical aspects of fibre spinning process in molten polymers. Part III. Tensile force and stress.* Koll.-Z., 175 (1961), 14-27. *(pp.94,102)*

Z9 A. Ziabicki, *Fundamentals of fibre formation. The science of fibre spinning and drawing.* Wiley-Interscience, London (1976). *(pp.12,16,17,94,95,97,99,213)*

Z10 M. Zidan, *On the rheology of the spinning process.* Rheol. Acta, 8 (1969), 89-123. *(pp.10,18,101,104)*

Subject index